THE HISTORY OF MODERN PHYSICS

BIBLIOGRAPHIES OF THE HISTORY
OF SCIENCE AND TECHNOLOGY
(Vol. 4)

GARLAND REFERENCE LIBRARY
OF THE HUMANITIES
(Vol. 420)

Volume 4

Bibliographies of the History of Science and Technology

Editors

Robert Multhauf
Smithsonian Institution, Washington, D.C.

Ellen Wells
Smithsonian Institution, Washington, D.C.

THE HISTORY OF MODERN PHYSICS
An International Bibliography

Stephen G. Brush
Lanfranco Belloni

GARLAND PUBLISHING, INC. • NEW YORK & LONDON
1983

Library of Congress Cataloging in Publication Data

Brush, Stephen G.
The history of modern physics.

(Bibliographies of the history of science and technology ; v. 4)
(Garland reference library of the humanities ; v. 420)
Includes indexes.
1. Physics—History—Bibliography. I. Belloni, Lanfranco.
II. Title. III. Series. IV. Series: Garland reference library of the humanities ; v. 420.
Z7141.B78 1983 [QC7] 016.539'09 82-49291
ISBN 0-8240-9117-5

Cover design by Laurence Walczak

Printed on acid-free, 250-year-life paper
Manufactured in the United States of America

GENERAL INTRODUCTION

This bibliography is one of a series designed to guide the reader into the history of science and technology. Anyone interested in any of the components of this vast subject area is part of our intended audience, not only the student, but also the scientist interested in the history of his own field (or faced with the necessity of writing an "historical introduction") and the historian, amateur or professional. The latter will not find the bibliographies "exhaustive," although in some fields he may find them the only existing bibliographies. He will in any case not find one of those endless lists in which the important is lumped with the trivial, but rather a "critical" bibliography, largely annotated, and indexed to lead the reader quickly to the most important (or only existing) literature.

Inasmuch as everyone treasures bibliographies, it is surprising how few there are in this field. Justly treasured are George Sarton's *Guide to the History of Science* (Waltham, Mass., 1952; 316 pp.), Eugene S. Ferguson's *Bibliography of the History of Technology* (Cambridge, Mass., 1968; 347 pp.), François Russo's *Histoire des Sciences et des Techniques. Bibliographie* (Paris, 2nd ed., 1969; 214 pp.), and Magda Whitrow's *ISIS Cumulative Bibliography. A bibliography of the history of science* (London, 1971–; 2131 pp. as of 1976). But all are limited, even the latter, by the virtual impossibility of doing justice to any particular field in a bibliography of limited size and almost unlimited subject matter.

For various reasons, mostly bad, the average scholar prefers adding to the literature, rather than sorting it out. The editors are indebted to the scholars represented in this series for their willingness to expend the time and effort required to pursue the latter objective. Our aim has been to establish a general framework which will give some uniformity to the series, but otherwise to leave the format and contents to the author/compiler. We have urged that introductions be used for essays on "the state of the field," and that selectivity be exercised to limit the length of each volume to the economically practical.

Since the historical literature ranges from very large (e.g., medicine) to very small (chemical technology), some bibliographies will be limited to the most important writings while others will include modest "contributions" and even primary sources. The problem is to give useful guidance into a particular field—or subfield—and its solution is largely left to the author/compiler.

In general, topical volumes (e.g., chemistry) will deal with the subject since about 1700, leaving earlier literature to area or chronological volumes (e.g., medieval science); but here, too, the volumes will vary according to the judgment of the author. The topics are international, with a few exceptions (Greece and Rome, Eastern Asia, the United States), but the literature covered depends, of course, on the linguistic equipment of the author and his access to "exotic" literatures.

Robert Multhauf
Ellen Wells

Smithsonian Institution
Washington, D.C.

CONTENTS

INTRODUCTION

The purpose of a bibliography is twofold: first, to allow the researcher to find, quickly and accurately, specific information about a subject; second, to offer guidance to those users who are just beginning to look into the literature of the subject. The advanced researcher will demand that the literature be exhaustively listed and indexed, and will have little need for an introduction (other than to learn how the bibliography is organized) or for evaluations of the works listed; while the beginner will profit from the compiler's judgment in selecting the most appropriate items to read first. In this particular bibliography, we lean toward the first purpose, since there are already many books and articles surveying the development of modern physics and we assume that the user of this bibliography will already have some idea of the outlines of the subject. Rather than provide extensive annotations and evaluations of the major works, we have concentrated on locating publications that may not be known to most English-speaking historians and scientists, and on making it possible to retrieve information about less-famous physicists. However, we do offer in this introduction some suggestions about general works for those who are not already familiar with the subject.

"Modern Physics" is defined here as physics since the discovery of X-rays in 1895. Geophysics and astrophysics are not included (except for relativistic cosmology) since separate bibliographies on those topics will be published in this series. There are a total of 2073 items.

The coverage of this bibliography is strongly influenced by the existence of two other reference works: first the *Isis Cumulative Bibliography* edited by Magda Whitrow, based on the annual critical bibliographies of the history of science published from 1913 to 1965, and its continuation by John Neu including bibliographies published from 1966 to 1975; second, *Literature on the History of Physics in the 20th Century* compiled by J.L. Heilbron and Bruce R. Wheaton. We assume that both these important tools, denoted by CB and HW respectively, are available to the user of our bibliography, and we have therefore followed the general principle that items included in either or both of them will *not* be listed unless we give an annotation or listing of contents. Although this principle was not strictly followed in all cases, it has enabled us to reduce the size (and price) of our bibliography and has allowed us, conversely, to include many "obscure" items that would have to be left out of a more selective bibliography.

In the Biography section, we have followed HW's practice by using the symbol @ to indicate that an article on the person may be found in the *Dictionary of Scientific Biography*. The *DSB* is generally the best single source of an authoritative concise article on any scientist, and we assume that it is available to the reader. Scientists who died after 1970 are generally not included but they may be covered in the supplemental volumes now being planned. We also follow HW by indicating Nobel Laureates with the symbol $ followed by the year of the award; a large amount of popular, secondary literature is often available for such persons, and may be retrieved through standard library catalogs, the *Reader's Guide to Periodical Literature*, the *New York Times* index, etc. Finally, we use the letters HW and CB to indicate that additional publications about this person are listed in those works. (Our imitation of the format of HW is intended not only as flattery but also as a convenience to the user.)

Since HW was published less than two years before our own compilation was finished, one might question the need for another bibliography on the same subject. The answer is found in the major categories of literature which HW chose to omit, but which we wanted to include: (1) biographical articles about physicists active after the mid-1950s; (2) obituaries, especially of second-rank physicists—in our own research we have found that it is often extremely difficult and time-consuming to locate even the basic facts such as birth and death dates for such people; (3) anthologies of reprints or translations of "classic" papers on a particular topic by several authors. When annotated older works, items published in the last two years and earlier items overlooked by HW are added, one has a fairly substantial collection.

Books of the third category mentioned above are often listed by HW and CB by editor and title only, but we have indicated the authors of individual papers (except for very short extracts) and have included names of those authors in our index, so that one can retrieve reprints or translations of papers by that author. This will help to answer one of the questions frequently asked of historians and librarians: where can I find, say, Einstein's 1905 paper on the photoelectric effect, if the original is either not available in my library or I need an English translation?

Another kind of literature which is completely omitted by CB and HW, probably because it has begun to appear only in the last few years, is the "citation classic" published in the weekly *Current Contents* service of the Institute for Scientific Information. The author of a frequently-cited paper is asked to write a brief essay describing the circumstances of the research and writing of this paper and to explain why it was frequently cited; these essays provide valuable information for historians of contemporary science.

The literature on the science of any historical period is growing year by year because historians continue to publish the results of their research; in the case of 20th-century science, this kind of growth is compounded by the yearly increase in the amount of science to be studied. Thus any bibliography such as this one must necessarily fail to include the items of greatest interest to the user, namely, those which appeared *after* the compilation of the bibliography. We therefore urge the user to consult the more recent *Isis* Critical Bibliographies, available in all major libraries. Another source of such information is the *History of Physics Newsletter*, published by the Division of History of Physics of the American Physical Society, currently edited by one of us (S.G.B.), which publishes the authors' summaries of books, articles, and papers presented at meetings. The "summaries" section of the *Newsletter* will be an ongoing supplement to this bibliography.

Those interested in following current historical research on science should also become regular readers of the following journals: *Historical Studies in the Physical Sciences: Archive for History of Exact Sciences; Isis; The British Journal for the History of Science; Annals of Science; Centaurus; Historia Mathematica; Earth Sciences History; History of Science; Journal for the History of Astronomy; Physis; Revue d'Histoire des Sciences; Archives Internationales d'Histoire des Sciences; NTM;* and *Berichte zur Wissenschaftsgeschichte.* In addition several journals published for physicists and physics teachers often carry reviews and historical or biographical articles: *American Journal of Physics; The Physics Teacher; Physics Today; Physics Bulletin; Reviews of Modern Physics;* and *Soviet Physics Uspekhi.* Every science or physics department library should subscribe to all these periodicals.

Those who plan to do their own research on the history of 20th-century physics should become familiar with the resources of the Center for History of Physics at the American Institute of Physics headquarters in New York City. The Center maintains files on unpublished correspondence, notebooks, oral history interviews and other documents which are available at archives anywhere in the world and publishes newsletters and brochures with information about such holdings and new additions to them.

As promised, we now present a "guided tour" through the literature of 20th-century physics as presented in this bibliography, with recommendations of books and articles useful for the student or physicist who wants an overview of a topic; in many cases these are works not actually cited here because they are listed in HW and/or CB. Preference is given to English-language publications.

Among general works on the history of modern physics, Cecil Schneer's *Evolution of Physical Science* (item A103) is a good introductory textbook though it does not cover the developments of the last two or three decades. Spencer Weart's article (A137) provides a rapid survey of the events of the last 50 years. Gerald Holton's *Introduction to Concepts and Theories in Physical Science,* as revised by one of us (S.G.B.), presents an exposition of major theories of 20th-century physics in the context of earlier developments (A47). Several famous physicists have written popular books giving their own overview of the growth of the subject to which they have made major contributions, and these are worth reading despite occasional historical inaccuracies; outstanding representatives of this category are the books by Einstein & Infeld (item A30), Gamow (A34), Heisenberg (A38, A39), and Segre (A107). The scientist-turned-novelist C.P. Snow has written a valuable account based in part on his personal encounters with physicists (A119). Among the collections of scholarly papers we note the book edited by Charles Weiner (A140) and the two books by Holton (A46, A47). Finally, the anthologies of annotated sources compiled by Boorse & Motz (A13) and by Sambursky (A102) provide an effective way to become better acquainted with modern physics and its historical context.

Readers who intend to make a serious study of some aspect of 20th-century physics will need to be proficient in German; the anthology by Beaton and Bolton (A5) provides a good opportunity to polish up this skill by reading selected major physics articles (the editors provide helpful notes on difficult words).

The biography section begins with a list of works about several physicists, arranged by author. Barbara Cline's book (B7) is a good survey for students. As indicated earlier, the *Dictionary of Scientific Biography* (B13) is the most authoritative and comprehensive source now available, but since it is too expensive for most personal libraries, one might want to purchase the *Concise* one-volume edition of the *DSB* (1981) or a more modest work such as Isaac Asimov's *Biographical Encyclopedia of Science and Technology* (New York: Doubleday, 1964, 1972 and later editions) or Trevor I. Williams' *A Biographical Dictionary of Scientists* (New York: Wiley, 3rd ed. 1982).

The main part of the biography section is arranged by biographee; thus section BE includes works about physicists whose names begin with the letter E. Under each biographee we have also listed autobiographies, editions of correspondence, and collected works. A well-known example of the former is Einstein's "Autobiographical Notes" in *Albert Einstein Philosopher-Scientist,* edited by P.A. Schilpp (Library of Living Philosophers, 1949; reprinted, New York: Harper, 1959), pp. 1-95. Articles dealing primarily with the person's contributions to a particular area of physics will be found under the section devoted to that area; thus one should use the index to locate all items about a person.

Of general interest in the section on social and institutional history are the books by Daniel Kevles on the American physics community (item C59) and by Alvin Weinberg on "Big Science" (C115). Also recommended are Charles Wiener's article on the transfer of physics and physicists from Europe to America in the 1930s (C16), and the study by Harriet Zuckerman and Robert K. Merton on the referee system as used by *Physical Review* (C120). The monograph by Forman, Heilbron, and Weart (C33) is a frequently-cited pioneering study of the physics community in 1900.

"Mechanics," defined so as to exclude quantum theory and relativity, has been of little interest to research physicists in the 20th century; a good survey is Truesdell's article (D9). Goldstein's reflections on the writing of his very popular textbook provide a good example of a "citation classic" (D2). We classify research on the mechanical and thermal properties of matter under this heading, however, so the following sections are more substantial. Ryogo Kubo, a leading theorist, gives a survey of statistical mechanics (DA8); it is placed in the section "Properties of Bulk Matter" (DA) because he includes transport properties as well as equilibrium theory. A more detailed account is in the forthcoming book by S.G. Brush, *Statistical Physics and the Atomic Theory of Matter from Boyle and Newton to Landau and Onsager* (Princeton, NJ: Princeton University Press, 1983), which includes the text of the article "Statistical Mechanics and the Philosophy of Science" (DB10) and covers the theory of phase transitions, quantum statistics, and the superfluidity of helium. Sydney Chapman's reminiscences (DC5) deal in part with the circumstances of his own contributions to transport theory; for further details, see item DC3.

The theory of Brownian movement, and its experimental confirmation which finally persuaded skeptics that matter does have an atomic structure, are reviewed in item DD2; see also the biography of Jean Perrin by Mary Jo Nye, *Molecular Reality* (New York: American Elsevier, 1972). For the history of fluid mechanics one should always consult the lively writings of C. Truesdell (e.g., DE8), whether or not one wants to accept his interpretations. Robert Beyer's article (DF1) is a brief survey of the recent history of acoustics. The Royal Society Symposium Proceedings

(DG54) include several excellent papers on the history of solid state physics; J.C. Slater provides a more personal account in his autobiography (BS15). R.P. Hudson offers a review of low-temperature physics (DH14), which may be followed by Kurt Mendelssohn's book *The Quest for Absolute Zero* (New York: Wiley, 1977). We did not find any comparable short review of high-temperature physics for section DI, but the new book on *Fusion* by Joan Bromberg (Cambridge, MA: MIT Press, 1982) and Bostick's article (DI1) cover the recent history of applied plasma physics.

Another classical subject, electricity and magnetism, would seem to have been rather neglected in this century, to judge by the size of section E; but additional items may be found in the sections on solid state physics (DG) and electromagnetic waves (G). J.H. Van Vleck's article on magnetism (E22) is a good introduction to the subject.

Section F, Relativity, brings us to one of the "revolutions" of 20th-century physics. Arthur I. Miller gives a comprehensive account of Einstein's 1905 special theory and its early interpretations (item FA19). Gerald Holton's essays in his *Thematic Origins* (A46) address various topics such as the alleged role of the Michelson-Morley experiment in Einstein's thinking. Swenson's *Genesis of Relativity* (FA34) discusses the relevant 19th-century background; Stanley Goldberg's article on "The Lorentz Theory of Electrons and Einstein's Theory of Relativity," *American Journal of Physics,* 37 (1969): 982-94 is also recommended. On general relativity, one might start with a sketch by the astrophysicist S. Chandreasekhar (FB34) and then read J. Mehra's discussion of the theories of Einstein and David Hilbert (FB22). C.W.F. Everitt reviews various experiments on gravity including his own (FC6), while G.J. Whitrow gives a rapid survey of the history of gravitational theory (FC12). Section FD lists works on several aspects of the response to relativity theory; a good introduction is the two-part article by Jeffrey Crelinsten, "Physicists Receive Relativity: Revolution and Reaction" and "Einstein, Relativity and the Press: The Myth of Incomprehensibility," *The Physics Teacher,* 18 (1980): 187-93, 115-22. Biographical works on Einstein (section BE) may also be consulted for all aspects of the history of relativity.

The recent history of optics is reviewed by Peter Franken (item G14). In the absence of a comprehensive survey of optics and electromagnetic waves in the 20th century, one might get a bird's-eye view by reading the addresses of those who received the Nobel Prize for work in this area: H.A. Lorentz (G28), P. Zeeman (G48), C.F. Braun (G4), G. Marconi (G31), A.A. Michelson (G33), P.A. Čerenkov (G7), I.M. Frank (G13), I.E. Tamm (G13), all collected in the series of *Nobel Lectures* (A87), and the recent addresses of N. Bloembergen (G1) and A.L. Schawlow (G39).

Quantum theory, the other major revolution in 20th-century physics, occupies sections H through HF and several of the works on Niels Bohr, Max Born, P.A.M. Dirac, Einstein, Werner Heisenberg, Max Planck and Erwin Schroedinger in the biography section. Dirac's article "The Evolution of the Physicist's Picture of Nature" (HD17) is an excellent overview as well as being itself a document of historical value. Victor Guillemin's *The Story of Quantum Mechanics* (H14) is a good elementary introduction to the entire subject, including some of the philosophical issues and the beginnings of elementary particle theory. Max Jammer's *Conceptual Development of Quantum Mechanics* (H20) goes into more technical detail and is an essential reference work. For scholars planning to undertake serious research on any topic related to the history of quantum theory, the guide to the Quantum Physics Archive prepared by Thomas S. Kuhn et al. (H2) is indispensable. (The Archive itself may now be consulted in half a dozen locations.) See also J.L. Heilbron, "Quantum Historiography and the Archive for History of Quantum Physics," *History of Science,* 7 (1968): 90-111.

The traditional view that Max Planck introduced the quantum hypothesis is presented by Martin J. Klein (HA20) and others; it has been challenged by Thomas Kuhn (HA22). Peter Galison (HA9) reviews the controversy. J.L. Heilbron gives a summary of the development of the Rutherford-Bohr atomic model in item HB20. For the invention of matrix mechanics one may consult Heisenberg's own account (HC30) as well as the relevant chapters in Jammer's book (H20). Two interesting articles on the development of wave mechanics are: Martin J. Klein, "Einstein and the Wave-Particle Duality," *The Natural Philosopher,* 3 (1964): 1–49, and Paul Forman, "Why was it Schroedinger who developed de Broglie's ideas?" *Historical Studies in the Physical Sciences,* 1 (1969): 291-314.

Anyone who is concerned with the philosophy of modern physics should read Niels Bohr's account of his "Discussion with Einstein on epistemological problems in Atomic Physics," in *Albert Einstein Philosopher-Scientist* edited by P.A. Schilpp, pp. 199–241. Abner Shimony's article, "Metaphysical Problems in the Foundations of Quantum Mechanics," *International Philosophical Quarterly,* 18 (1978): 3–17, is a fascinating summary of the current status of the question argued by Bohr and Einstein: Does the world have objective reality apart from our observation of it? Max Jammer's *Philosophy of Quantum Mechanics* (HD36) is the definitive work on the history of this subject. Paul Forman's article on the effect of Weimar culture on the acceptance of indeterminism (HD22) has become a classic example of the claim that physical theories can be influenced by the social environment; see also John Hendry's critique of it (HD29). On the history of quantum field theory and quantum electrodynamics, start with V.F. Weisskopf's article (HF8) or Steven Weinberg's "The Search for Unity: Notes for a History of Quantum Field Theory," *Daedalus,* 106 (1977): 17–35.

Sections JA through JH cover research on X-rays, elementary particles and nuclear physics. One might begin with Bern Dibner's short book, *Wilhelm Conrad Roentgen and the Discovery of X-Rays* (New York: Watts, 1968) and then read David L. Anderson's *The Discovery of the Electron* (Princeton, NJ: Van Nostrand, 1964). Robert A. Millikan's famous oil-drop experiment has recently been the subject of critical analysis by Gerald Holton (item JB11) and Allan Franklin (JB10). The dramatic story of the discovery of radioactivity and nuclear transmutation is told in biographies of Marie Curie and Ernest Rutherford (e.g., E.N. da C. Andrade, *Rutherford and the Nature of the Atom,* Garden City, NY: Doubleday/Anchor, 1964); for an example of a more scholarly approach see Lawrence Badash, "'Chance Favors the Prepared Mind': Henri Becquerel and the Discovery of Radioactivity," *Archives Internationale d'Histoire des Sciences* 70–71 (1965): 55–66. An authoritative account of Rutherford and Soddy's discovery of transmutation is Thaddeus Trenn's monograph, *The Self-Splitting Atom* (JC46). George Trigg's *Crucial Experiments in Modern Physics* (JD20) is a useful introduction to several aspects of atomic physics before 1930. Recent attempts to test quantum mechanics against the possibility of "hidden variables" are examined from a sociological angle by Bill Harvey (JE12, JE13).

Of the many publications on the history of nuclear fission and the beginnings of elementary particle research in the 1930s, David L. Anderson's chapters prepared for the Project Physics Course (item JF5) may be noted as especially suitable for students, together with Hans Graetzer's article "Discovery of Nuclear Fission," *American Journal of Physics,* 32 (1964): 9–15. Another useful article is Charles Weiner, "1932—Moving into the New Physics," *Physics Today,* 25, no. 5 (May 1972): 40–49. On the history of elementary particles, see Laurie M. Brown and Lillian Hoddeson, "The Birth of Elementary-Particle Physics" (JF10) and V. Mukherji, "A Short History of the Meson Theory from 1935 to 1943," *Indian Journal of History of Science,* 6 (1971): 75–101, 117–34. A piece of very recent history is presented in Andrew Pickering's study of "The Hunting of the Quark" (JH45). An important source for recollections of scientists is the proceedings volume of the *Colloque International sur l'Histoire de la Physique des Particules: Quelques decouvertes, concepts, institutions des annees 30 aux annees 50* (Paris: Editions de Physique, 1982).

The interaction between physics and philosophy (section K) is such a broad subject that no single book can do justice to it. Phillip Frank's *Modern Science and its Philosophy* (K48) is a good introduction; this might be followed by Niels Bohr's presentation of his views in *Atomic Theory and the Description of Nature* (K5) and David Bohm's *Casuality and Chance in Modern Physics* (K4). Similarly, the cultural influences of physics have generated a large literature; a good example of recent scholarship is Robert Nadeau's book on physics and metaphysics in the novel (L23). The analogy between theories of modern physics and ideas of Eastern philosophy, suggested by Fritjof Capra in his popular book *The Tao of Physics* (Berkeley, 1975), is examined in Sal Restivo's article (L25).

In the last section (M) we collect some publications on research in the history of modern physics and its use in education. The conference proceedings edited by S.G. Brush and A.L. King (item M5) summarizes a variety of activities.

ACKNOWLEDGMENTS

Preparation of this bibliography was supported in part by a grant from the National Endowment for the Humanities. We thank the following scholars who contributed several items: Dr. Viktor Iakovlevich Frenkel' (Ioffe Physico-Technical Institute of the USSR Academy of Sciences, Leningrad); Dr. Vladimir Pavlovich Vizgin and V.K. Poltavets (Institut Istorii Estestvoznaniia i Tekhniki of the USSR Academy of Sciences, Moscow); Dr. Jozseph Illy (Institute of Isotopes of the Hungarian Academy of Sciences, Budapest); Dr. Gabriel Strempel and Nicholas Burghelea (Biblioteca of Academia Republicii Socialiste Romania); Prof. Lubos Noby (Department of History of Science and Technology of the Institute of Czechoslovak and General History of CSAS); N.H. Robinson (Librarian, The Royal Society, London); Dr. Alfred Guenther (Scientific Information Service, CERN, Geneva); several members of the Lehrstuhl für Geschichte der Naturwissenschaften and Technik, Universitaet Stuttgart, Federal Republic of Germany; Dr. J. Teichmann (German Team of the International Project for History of Solid State Physics, Deutsches Museum, Munich, Federal Republic of Germany). For assistance in preparing the material for publication we are indebted to Nicholas Brush and Martin Collins.

THE HISTORY OF MODERN PHYSICS

A. HISTORY OF MODERN PHYSICS -- GENERAL & MISCELLANEOUS WORKS

A1 Agranovskiĭ, V. A. Vziâtie Sto Chetvertogo. Moskva: Znanie, 1967. 108 pp.

Essays on the work of Soviet physicists.

A2 American Institute of Physics, Center for the History of Physics. "Two Hundred Years of American Physics." Physics Today, 29, no. 7 (July 1976): 23-31.

A3 Artsimovich, L. A., Editor. Razvitie Fiziki v SSSR. 2 volumes. (A. N. SSSR, Otdelenie Obshcheĭ i Prikladnoĭ Fiziki. Institut Istorii Estestvoznaniia i Tekhniki) Moskva: Nauka, 1967. xv + 451; 363 pp.

The development of physics in the USSR.

A4 Auerbach, Felix. Entwicklungsgeschichte der modernen Physik, Zugleich eine Uebersicht ihrer Tatsachen, Gesetze und Theorien. Berlin: Springer, 1923. viii + 344 pp.

Primarily an exposition of quantum theory and statistical mechanics.

A5 Beaton, K. B., and H. C. Bolton, eds. A German Source-Book in Physics. New York: Oxford University Press, 1969. xvi + 315 pp.

Anthology of German papers in 20th century physics, for English speaking students who are learning German.

A6 Bedreag, C. C.: Bibliografia fizicii române. Biografii. Bucuresti: Ed. Tehnică, 1957. 293 pp.

A7 Belloni, Lanfranco. "Scienze Fisiche 1970-1980." Annali della Scienza e della Tecnica 1980, Enciclopedia della Scienza e della Tecnica. Milano: Mondadori, 1981, pp. 796-802.

A8 Bernstein, Jeremy. A Comprehensible World: On modern Science and its Origins. New York: Random House, 1967. xii + 269 pp.

Includes essays on T.D. Lee & C. N. Yang (discovery of parity nonconservation), CERN, Erwin Schroedinger, and the Einstein-Bohr debate.

A9 Blin-Stoyle, R.J., et al. Turning Points in Physics. Amsterdam: North-Holland, 1959; New York: Harper & Row, 1961. 192 pp.

Lectures at Oxford in 1958, by R. J. Blin-Stoyle, D. ter

Haar, K. Mendelssohn, G. Temple, F. Waismann and D. H. Wilkinson, on field theory, relativity and elementary particles.

A10 Bloch, Leon. "Les Theories newtoniennes et la Physique moderne." Revue de metaphysique et de morale, 35 (1928): 41-54.

A11 Bloembergen, N. "Physics." Science in Contemporary China (item A90), pp. 85-109.

A discussion of recent history, mostly institutional.

A12 Bohm, David. "Quantum theory as an indication of a new order in physics. Part A: The development of new orders as shown through the history of physics." Foundations of physics, 1 (1971): 359-81.

A13 Boorse, Henry A., and Lloyd Motz, eds. The World of the Atom. New York: Basic Books, 1966. xxvi + 1873 pp.

Anthology with selections from papers on radioactivity, quantum theory, and nuclear physics.

A14 Bothe, Walther. Der Physiker und sein Werkzeug. Berlin: W. de Gruyter, 1944. 26 pp. (Preussische Akademie der Wissenschaften, Vorträge und Schriften, Heft 22)

A15 Brown, Sanborn C., ed. Physics 50 Years Later, as presented to the XIV General Assembly of the International Union of Pure and Applied Physics on the occasion of the Union's Fiftieth Anniversary, September 1972. Washington, D.C.: National Academy of Sciences, 1973. vii + 406 pp., illustrations.

A collection of articles assessing progress in various areas of physics, by E. Amaldi, J. T. Wilson, F. Hoyle, G. Toraldo di Francia, G. Herzberg, J. Bardeen, J. Friedel, R. W. Gould, A. Bohr & B. R. Mottelson, V. F. Weisskopf, K. S. Thorne, H.G. Stever, D. A. Bromley, W. Gentner, H. B. G. Casimir and R. F. Bacher; includes documents from an exhibit, "Physics in 1922" prepared by J. N. Warnow, and an article "The International Union of Pure and Applied Physics from 1923 to 1972" by P. Fleury.

A16 Brown, Stanley B., ed. The New Science: Recent Advances in Physics. Louisvile, Ky.: Touchstone Pub. Co., 1972. 192 pp.

Reprints of articles on various topics in modern physics.

A17 Buckley, H. A Short History of Physics. New York: Van Nostrand, 1927. xi + 263 pp.

A18 Caldirola, Piero, and Angelo Loinger. _Teoria fisica e realtà_. Napoli: Liguori, 1979. 191 pp.

Contains the following semi-historical essays: "Il principio di esclusione," 51-65 (also in _Scientia_, 110 (1975): 51-67); "Boltzmann, Planck e il corpo nero," 93-112; "Storie e filosofie della fisica," 149-62; an essay by B. Bertotti on the epistemology of gravitational physics (item FC3); and a review by E. Fiorini on elementary particles.

A19 Carazza, B. "The History of the Random-Walk Problem: Considerations on the Interdisciplinarity in Modern Physics." _Rivista del Nuovo Cimento_, serie 2, 7(1977): 419-427.

A20 Cavallo, Giacomo and Antonio Messina. "Caratteri, ambienti e sviluppo dell'indagine fisica nel Novecento e la politica della ricerca." _Storia d'Italia_ (item A82), pp. 1109-62.

Survey of Italian physics in the 20th century.

A21 Chase, Carl Trueblood. _The Evolution of Modern Physics_. New York: Van Nostrand, 1947.

A22 Christman, Ruth C., ed. _Soviet Science_. Washington: American Association for the Advancement of Science, 1952. 108 pp.

Proceedings of a symposium. Contains items A128 & A150.

A23 Chwolson, O. D. _Die Physik 1914-1926. Siebzehn ausgewaehlte Kapitel_. Aus dem Russischen uebersetzt von Georg Kluge. Braunscwheig: Vieweg, 1927. ix + 696 pp.

Chapters on the electron, quanta, atomic structure, line spectra, ultraviolet and infrared rays, excitation and ionisation of gases by electron collisions (Franck-Hertz experiment), photoelectric efect, the Bohr model, isotopes, superconductors, Michelson's new experiment (1925), electromagnetic spectroscopy of metals.

Note: This author is listed as "Khvolson, Orest Daniilovich" in many catalogs, including HW.

A24 Cohen, E. Richard. "This Week's Citation Classic. Cohen E R & DuMond J W M. Our knowledge of the fundamental constants of physics and chemistry in 1965. _Rev. Mod. Phys_. 37: 537-94, 1965." _Current Contents, Physical, Chemical & Earth Sciences_, 21, no. 35 (August 31, 1981): 16.

The author recalls the circumstances of writing this

paper, which has been cited more than 310 times, and comments on subsequent reviews of this topic.

A25 Danin, D. S. Neizbezhnost' Strannogo Mira. 3rd Edition. Moskva: Molodaîa Guardiîa, 1966. 375 pp.

Popular-scientific book about physics and physicists.

A26 Dorfman, Iakov G. Vsemirnaîa Istoriîa Fiziki (s Nachala XIX do Serediny XX). Moskva: Nauka, 1979. 317 pp.

World history of physics from the beginning of the 19th century to the first half of the 20th century.

A27 Dubinsky, Juraj. "Looking back upon the 35 years and beyond." Acta physica Slovaca. 39, nr. 2 (1980): 105-6.

A28 Dugas, Rene. A History of Mechanics. Neuchatel: Editions du Griffon, 1955. 671 pp.

Includes chapters on relativity and quantum theory.

A29 Dyson, Freeman J. "Missed opportunities." Bulletin of the American Mathematical Society, 78 (1972): 635-52.

Examples in which physicists and mathematicians missed discoveries through failure to talk to each other.

A30 Einstein, Albert, and Leopold Infeld. The Evolution of Physics. New York: Simon and Schuster, 1938. x + 320 pp.

Mostly on relativity theory and its 19th-century background.

A31 Feuer, Lewis S. Einstein and the Generations of Science. New York: Basic Books, 1974. ix + 374 pp.

Includes sections on "The Social Roots of Einstein's Theory of Relativity," "Social, Generational,and Philosophical Sources of Quantum Theory," and generational movements in science.

A32 Frenkel, Victor Ia., "O zhanre biografii uchenykh." Chelovek Nauki. (Akademiia Nauk SSSR, Institut Istorii Estestvoznaniia i Tekhniki) Moskva: Izdatel'stvo "Nauka", 1974, pp. 108-24.

Discusses briefly the problems of scientific biography.

A33 Frenkel, V. Ia., and E. V. Shpol'skii. "Fizicheskie nauki." Bolshaîa Sovetskaîa Entsiklopediîa. 3rd edition. T. 24, vol. 2. Moskva: 1977, pp. 856-872.

A33 Friedman, Robert Marc. "Nobel Physics Prize in Perspective." *Nature*, 292 (1981): 793-98.

"The history of the award of the Nobel Prize in Physics, which is partly the history of interpreting the statutes, also depends on the development of physics in Sweden. The now traditional restricted scope of the prize has its origin in past controversies, not formal rules. Einstein's relativity and Bjerknes's meteorology were both casualties." (Author's abstract) The award to Einstein was delayed by the opposition of A. Gullstrand; C. W. Oseen succeeded in blocking any award to V. Bjerknes and in excluding geophysics from the scope of the Physics prize.

A34 Gamow, George. *Biography of Physics*. New York: Harper, 1961. viii + 338 pp. Translation: *Biografia fizicii* (translated by M. Sabău & T. Rosescu) Bucureşti: Ed. Ştiinţifica, 1971. 386 pp.

The second half of the book is on relativity, quantum theory, the nucleus and elementary particles.

A35 Gans, Richard. *Die Physik der letzten dreissig Jahre*. Koenigsberg: Grafe und Unzer, 1930. 19 pp.

A36 Gould, Sydney, ed. *Sciences in Communist China*, A symposium presented at the New York meeting of the American Association for the Advancement of Science, December 26-27 1960. Publication no. 68. Washington, D. C.: American Association for the Advancement of Science, 1961. 872 pp.

Contains T. Y. Wu, "Nuclear Physics," 631-643; Robert T. Beyer, "Solid state physics," 645-658.

A37 Halbwachs, F. "Structure de quelques révolutions scientifiques en physique." *Raison Présente*, 32 (1974): 85-101.

Application of the ideas of Marx, Piaget and Kuhn to wave vs. particle theories of light, special relativity and quantum mechanics.

A38 Heisenberg, Werner. *Philosophic Problems of Nuclear Science*. Translated by F. C. Hayes. New York: Fawcett World Library, 1952. 144 pp.

Essays on recent changes in the foundations of physical science.

A39 Heisenberg, Werner. _Across the Frontiers_. Translated from German by Peter Heath. New York: Harper & Row, 1974. xxii + 229 pp.

Essays on A. Einstein, M. Planck, W. Pauli, philosophical problems of atomic theory, etc.

A40 Heisenberg, Werner. _Wandlungen in den Grundlagen der Naturwissenschaft_. Die 11. Auflage 1980 mit einem Beitrag "Werner Heisenberg und die moderne Physik" von G. Rasche und B.L. van der Waerden. Stuttgart: Hirzel Verlag, 1981. xxxi + 183 pp.

* Hermann, Armin, ed. _Lexikon Geschichte der Physik A-Z._ Reviewed by F. Krafft, cited below as item M39.

A41 Hermann, Armin, and Ulrich Benz. "Quanten- und Relativitätstheorie im Spiegel der Naturforscherversammlungen 1906-1920." _Wege der Naturforschung 1822-1972 im Spiegel der Versammlungen Deutscher Naturforscher und Artze_. Edited by H. Querner and H. Schipperges. Berlin, 1972, pp. 125-37.

A42 Hermann, A., and Friedrich R. Wollmershaeuser. "Die Entwicklung der Physik." _Festchrift zum 150 jaehrigen Bestehen der Universitaet Stuttgart_. Edited by Johannes H. Voight. Stuttgart, 1979, pp. 241-76.

A43 Hermann, A. _Weltreich der Physik Von Galilei bis Heisenberg_. Esslingen und Munchen: 1980. 392 pp.

A44 Hesse, Mary B. _Forces and Fields: A study of Action at a Distance in the History of Physics_. Totowa, NJ: Littlefield, Adams, 1961. x + 318 pp.

Includes two chapters on forces in modern physical theory.

A45 _Highlights of British Science. Based on the subjects of exhibits arranged for the Jubilee Exhibition at the Royal Society, 20-25 June 1977_. London: The Royal Society, 1978. 240 pp.

Includes: R. M. Longstaff, "Science and the development of nuclear energy"; S. J. Robinson, "The Jubilant electron"; M. L. Jenkins and M. J. Whelan, "Developments in electron microscopy and microanalysis."

A46 Holton, Gerald. _Thematic Origins of Scientific Thought: Kepler to Einstein_. Cambridge, Mass.: Harvard University Press, 1973. 495 pp.

Includes revised & expanded versions of items HD31 and K59.

A47 Holton, Gerald, and Stephen G. Brush. _Introduction to concepts and theories in Physical Science_. Second edition. Reading, Mass.: Addison-Wesley, 1973. xix + 589 pp.

First published by Holton in 1952; the new edition has additional material on 20th-century physics.

A48 Holton, Gerald. _The Scientific Imagination: Case Studies_. New York: Cambridge University Press, 1978. xvi + 382 pp.

Includes revised & expanded version of item JB11.

A49 Holton, Gerald, and Katherine Sopka. "Great books of science in the twentieth century: Physics." _Great Ideas Today_, Encyclopedia Britannica, Inc. (1979), pp. 225-77.

A50 Hull, Gordon Ferrie. "Fifty years of physics--a study in contrasts." _Science_, 104 (1946): 238-44.

* _Isis Cumulative Bibliography_, cited below as item B46.

A51 Ivanenko, D. D. "Problemy Istorii Frantsuzskoǐ Teoreticheskoǐ Fiziki." _Tezisy Dikladov i Soobshchenii na Mezhvuzovskoǐ Konferentsii po Istori Fizico-Matematicheskikh Nauk_. Moskva, 1960, pp. 92-93.

On the history of French theoretical physics since the beginning of the 19th century.

A52 Ivanenko, D. D. "Frantsuzkaîa Shkola Teoreticheskoǐ Fizike." _Iz Istorii Frantsuzskoi Nauki_ (Sbornik Statei). Moskva, 1960, pp. 156-181.

On the French school of theoretical physics.

A53 Jacob, M., Ed. _CERN, 25 Years of Physics_. Physics Reports Reprint Book Series, vol. 4. New York: North-Holland Pub. Co., 1981. 568 pp.

A54 Jordan, Pascual. _Das Bild der modernen Physik_. Hamburg-Bergedorf: Stromverlag, 1947.

A55 Jordan, Pascual. _L'immagine della fisica moderna_. Milano: Feltrinelli, 1964. 137 pp.

"Popularization" of modern quantum physics, its significance and its relations with biology.

A56 Kangro, Hans. "Die Bedeutung des Energieerhaltungsgesetzes für die Physik von ca. 1920-1932." _35. Physikertagung 1970, Hannover_. Edited by E. Schmutzer. Stuttgart: Vorabdruck der Fachberichte, 1970, pp. 191-95.

A57 Kedrov, B. M., and N. F. Ovchinnikov, eds. _Printsip Simmetrii: Istoriko-Metodologicheskie Problemy_. Moskva: Nauka, 1978. 397 pp.

The principle of symmetry as an historico-methodological problem.

A58 Kedrov, Bonifatij M. "On scientific revolutions and their typology. Part II: The last two types." _Scientia_, 115 (1980): 5-21. Italian version, _ibid_: 23-36.

Classification of the revolution in physics in the early 20th century as "third type" from Marxist viewpoint.

* Khvolson, O. D. _See_ Chwolson, O. D.

A58 Kleinert, Andreas. "Albert Einstein, Otto Hahn, Max von Laue, Lise Meitner. Zum 100. Geburtstag der Begrunder des Atomzeitalters." _Buch und Bibliothek_, 32 (1980): 507-17.

A59 Koizumi, Kenkichiro. _The Development of Physics in Meiji Japan: 1868-1912_. Ph. D. Dissertation, University of Pennsylvania, 1973. 295 pp.

For summary see _Dissertation Abstracts International_, 34 (1973): 1828-A.

A60 Konya, Albert. "Physics." _Science and Scholarship in Hungary_. Edited by T. Erdey-Gruz and K. Kulcsar. Second Edition. Corvina Press, 1975, pp. 40-59.

Survey of research in physics in Hungary in the last few decades, arranged by topic: particle physics, nuclear physics, atomic and molecular physics, solid state, other areas, teaching.

A61 Kravets, T. P. _Ot N'iutona do Vavilova: Ocherki i Vospominiia_. Leningrad: Nauka, 1967. 447 pp.

From Newton to Vavilov: essays and recollections. Contains chapters on Russian physicists active in the period from the end of 19th century to the first half of the 20th century, and on P.Lebedev and the discovery of radiation pressure.

A62 Kuznetsov, B. G., ed. _Razvitie sovremennoi fiziki_. Moscow: Izdatel'stvo "Nauka", 1964. 331 pp.

A collection of articles on the development of modern physics which includes several on the work of Niels Bohr.

A63 Kuznetsov, B. G. *Ocherki Fizicheskoĭ Atomistiki XX Veka.* Moskva: Nauka, 1966. 192 pp.

Essays on physical atomistics of the 20th century.

A64 Kuznetsov, B. C. *Von Galilei bis Einstein: Entwicklung der physikalischen Ideen.* Edited by G. Buchheim and S. Wollgast. Berlin: Akademie-Verlag, 1970. Basel: C.F. Winter, 1970. 391 pp.

A65 Kuznetsov, B. G. "Galiléi et Einstein: Prologue et épilogue de la science classique." *Actes, XIIe Congres International d'histoire des sciences, 1968*, 5 (1971): 59-63.

A66 Langevin, Paul. *Gîndire şi acţiune.* Texte culese şi prezentate de Paul Laberenne. In româneşte de Carol Neumann şi Cornelia Papacostea. Cuvînt înainte de prof. Valer Novacu. Bucureşti: Ed. Ştiinţifică, 1961. 259 pp.

A67 Laue, Max von. *Geschichte der Physik.* 2 Auflage. Bonn: Universitäts-Verlage, 1947. 148 pp. Translation: *Istoria Fizicii* (translated by I. Gornstein) Bucureşti: Ed. Stiintifica, 1965. 268 pp.

A68 Lenaizan, Beaulard de. "Sur l'evolution de loi physique." *Revue International de l'Enseignement*, 76 (1922): 146-61; 193-205.

A69 Leshkovtsev, V. A. *50 Let Sovetskoi Fiziki.* (Preface by L. A. Artsimovich) Moskva: Znanie, 1968. 47 pp. (Novoe v Zhizni, Nauke, Tekhnike)

50 years of Soviet physics.

A70 Loeb, Leonard B., and Arthur S. Adams. *The Development of Physical Thought.* New York: Wiley, 1933.

A71 Lupasco, Stephane. *La Tragedie de l'Energie: Philosophie et sciences du XXe Siecle.* Tournai: Casterman, 1970.

A72 Macorini, E., ed. *Annali della Scienza e della Tecnica Contemporanee, 1875-1975*. Part of the series *Scienziati e tecnologi contemporanei*. Milano : Arnoldo Mondadori, 1974, Vol. III, pp. 209-738.

"Sinossi" (synopses) of the development of the experimental and deductive sciences from 1875-1975. Essentially a chronology or a chrono-history of the different sciences "in parallel" divided in sub-periods: 1875-1900, 1900-1910,1910-1920,..., 1960-1970, 1970-1975. Piero Caldirola and Erasmo Recami eds. for "Scienze Fisiche." The list of authors and subjects includes: Arecchi F. Tito (Elettronica Quantistica), M. Baldo (Stati della Materia), E. Bellone (Scienze Fisiche 1875-1900) P. Caldirola(Teorie Fondamentali, Componenti fondamentali della materia, Fisica dei Plasmi), T. Chersi (Scienze Fisiche 1970-1975), E. Fiorini (Raggi cosmici), E. Rimini (Stati della materia), A. Mancini Nunzio (Proprieta della materia), E.Recami (Teorie fondamentali e componenti fondamentali della materia), E. Sindoni(fisica del plasma).

A73 Macorini, E., ed. *Scienze e technica del Novecento*, vol. I. *Annali dal 1900 al 1950*. (Biblioteca della EST, Edizioni Scientifiche e Tecniche) Milano: Mondadori, 1977, 334 pp. *Scienza e tecnica del Novecento*, vol. II, *Annali dal 1950 a oggi*, Edited by E. Macorini (Biblioteca della EST, Edizioni Scientifiche e Tecniche) Milano: Mondadori, 1977, pp. 336-610.

Same item as the preceding; *Sinossi* slightly updated to the mid-seventies. That is, the Sinossi of different sciences published in *Scienziati e tecnologi contemporanei*, vol. e, p.209-738, have been reissued separately in two volumes. The presentation is the same: a chronohistory of parallel developments divided in decades. The period 1875-1900 has been dropped and it starts with 1900-1910, 1910-1920, ..., 1960-1970, the seventies. P. Caldirola and E. Recami are editors for Science Fisiche. The authors are the same as in preceding item.

A74 Maglich, Bogdan, ed. _Adventures in Experimental Physics. A Selection of Papers and Personal Discovery Stories dealing with innovative, unconventional, and adventurous experimentation_. Princeton, NJ: World Science Communications, 1972-76. 5 vols. 360 + 168 + 172 + 176 pp.

Contents of "alpha volume": First optical pulsar, quantized circulation in superfluid, measurement of nuclear reaction time using "blocking effect," muon-induced fission, two kinds of neutrinos, transition radiation from ultrarelativistic particles, Apollo 11 Laser Ranging Retro-Reflector, search for hidden chambers in the Pyramids using cosmic rays.

Beta volume: Neutron, Maffei galaxies, Mendelevium, collisions of high energy protons with supersonic stream of hydrogen, search for quarks, discovery of antideuteron, bomb-propeled spaceship models.

Gamma volume: synthesizing Be^8 and effect of atomic electrons on nuclear resonances, Josephson effect and measurement of e/h, parity violation in weak interactions.

Delta volume: First chemical analysis of the lunar surface, positronium, electron tunneling into superconductors.

Epsilon volume: Transistor effect, first fusion weapons from a thermonuclear weapon device, omega meson -- first neutral vector meson, massive neutral vector mesons.

The publisher announced in 1982 that this series would be reissued and revised under the title _Adventures in Science_.

A75 Malisek, V. "Rozvoj fiziky v Ceskoslovensku za poslednich 30 let." _Matematika a fyzika ve škole_, 8 (1977-78): 674-78.

Czech physics in the last 30 years.

A76 Margenau, Henry. "Physics." _The Development of the Sciences, Second Series_. Edited by L. L. Woodruff. New Haven: Yale University Press, 1941, pp. 91-120.

A77 Marton, L., and C. Marton, eds. *Advances in Electronics and Electron Physics*. Vol. 50. New York-London: Academic Press, 1980. 493 pp.

M. Stanley Livingston, "Early History of Particle Accelerators," 2-88; Pierre Grivet, "Sixty Years of Electronics," 89-174; Charles Susskind, "Ferdinand Braun: Forgotten Forefather," 241-260; A. Van Der Ziel, "History of Noise Research" (analyzes early noise work of Einstein and Langevin), 351-409; James E. Brittain, "Power Electronics at General Electric: 1900-1941," 411-447; L. Marton and C. Marton, "Evolution of the Concept of Elementary Charge," 449-72.

A78 Mathieu, V., and P. Rossi, eds. *Scientific Culture in the Contemporary World*. Milano: Scientia, 1979. 420 pp.

Special volume published by the review *Scientia* in collaboration with UNESCO.

A79 Matsonashvili, B. N., ed. *V Glub Atoma.* Moskva: Nauka, 1964. 391 pp.

Reprint of articles previously published in *Priroda* on investigation of atom and nucleus by I. E. Tamm, Ia. A. Smorodinskiĭ, D. I. Blokhintsev, V. I. Gol'danskiĭ, I. M. Lifshits, and others.

A80 Mehra, Jagdish, ed. *The Physicist's Conception of Nature*. Boston: Reidel, 1974. xxiii + 839 pp.

Proceedings of a symposium honoring P. A. M. Dirac on his 70th birthday, held at Trieste, 1972.

A81 Merleau-Ponty, J., *Leçons sur la genèse des theories physiques. Galilee, Ampère, Einstein*. Paris: Vrin, 1974. 172 pp.

A82 Micheli, Gianni, ed. *Storia d'Italia. Annali 3. Scienza e Tecnica nella cultura e nella societa dal Rinascimento*. Torino: Einaudi, 1980. 1365 pp.

Includes: R. Maiocchi, "Il ruolo delle scienze nello sviluppo industriale italiano" and A20.

A83 Millikan, Robert A. *Evolution in Science and Religion*. New Haven: Yale University Press, 1927. Reprinted, with additions, Port Washington, N.Y.: Kennikat Press, 1973. 95 pp.

Chapter I: "The Evolution of Twentieth Century Physics."

A84 Moulton, ForestRay, and JustusJ. Schifferes. *The Autobiography of Science*. London: Murray, 1963. 748 pp.

A85 Needham, Joseph, and Walter Pagel, eds. _Background to Modern Science: Ten Lectures at Cambridge arranged by the History of Science Committee, 1936_. New York: Macmillan, 1938. xxii + 243 pp.

Includes: E. Rutherford, "Forty Years of Physics," (on the history of radioactivity and atomic structure), 49-54; W. L. Bragg, "Forty Years of Crystal Physics," 77-89 (item DG6); F. W. Aston, "Forty Years of Atomic Theory," 93-114.

A86 _Nobel Lectures, including Presentation Speeches and Laureates' Biographies. Chemistry_. Vol. 1, 1901-21; vol. 2, 1922-41; vol. 3, 1942-62; vol. 4, 1963-70. New York: Elsevier, 1964-72. xii + 411, xii + 508, xiii + 712, x + 359 pp.

A87 _Nobel Lectures, including Presentation Speeches and Laureates' Biographies. Physics_. Vol. 1, 1901-21; Vol. 2, 1922-41; vol. 3, 1942-62; vol. 4, 1963-70. New York: Elsevier, 1964-72. xii + 500; xii + 458; 621; xi + 351 pp.

A88 Nowak, K. "Rückschau ueber 20 Jahre." _Neue Physik_, 4 (1964): 12-32.

A89 Oppenheimer, J. Robert. "Perspectives in modern physics." _Perspectives in Modern Physics, Essays in Honor of Hans A. Bethe on the occasion of his 60th birthday_. Edited by R. E. Marshak. New York: Interscience, 1966, pp. 9-20.

A90 Orleans, Leo A., ed. _Science in Contemporary China_. Stanford, CA: Stanford University Pres, 1980. xxxii + 599 pp.

Includes item A11.

A91 Pais, Abraham, et al. "Twenty Years of Physics." _Physics Today_, 2, no. 5 (May 1968): 23-72.

A92 Polvani, Giovanni. "Fisica," _Un secolo di progresso_... (item A115), pp. 555-699.

Wide and sound historical survey of Italian physics from 1839 to 1939.

A93 Predvoditelev, A. S., and B. I. Spasski, eds. _Razvitie Fiziki v Rossii. Ocherki_. T. 2. _Ot Velikoi Oktiabr'skoi Sotsialisticheskoi Revoliutsii do Nashego Vremeni_. Moskva: Prosveshchenie, 1970. 447 pp.

The development of physics in Russia, from the Great October Socialist Revolution to about 1970. Contents include: P. S. Kudriavtsev on the founding of Soviet theoretical physics, lists of Soviet physicists who won the Lenin Prize or Nobel Prize, and articles on special topics.

A94 Ramsey, Norman F. "Physics in 1981 ± 50." Physics Today, 34, no. 11 (Nov. 1981): 26-34.

Lead article in an issue commemorating the 50th anniversary of the American Institute of Physics, preceded by a collection of photographs illustrating "Fifty Years of physics in America."

A95 Regge, Tullio. "La matematica nelle scienze naturali." Bollettino U. M. I., (5), 17-A (1980): 365-375.

Conference held at a meeting of the Italian Union of Mathematicians discussing the role of mathematics in recent theoretical physics at a technical level.

A96 Reichen, Charles-Albert. A History of Physics. New York: Hawthorn, 1963. 112 pp.

A97 Reĭnov, N. M. "Vospominanii͡a o tom, kak Delalis' Pribory." Khimii i Zhizn'. 9 (1970): 50-61, 69-70; 10 (1970): 37-50; 11 (1970): 72-86.

Recollection of Soviet physicists and chemists of the end of 1920s.

A98 Richtmyer, F. K. "The romance of the next decimal place." Science, 75 (1932): 1-5.

On increasing accuracy in physical measurements.

A99 Romer, Alfred. The Development of Atomic Physics. Canton, N.Y., 1949. ii + 75 pp.

A100 Rompe, Robert, and Hans-Jürgen Treder. "Die grosse Zeit der Physik in Berlin." Physiker über Physiker II, (item B25), pp. 9-25.

A101 Rousseau, Pierre. Survol de la Science Francaise Contemporaine. Paris: Fayard, 1974. 354 pp.

Survey of French science from 1939 to 1974.

A102 Sambursky,Shmuel, ed. Physical Thought from the Presocratics to the Quantum Physicists. An Anthology. New York: Pica Press, 1974. xv + 584 pp.

Includes selections from Planck, Einstein, Rutherford, Bohr, de Broglie, Heisenberg, and Pauli.

A103 Schneer, Cecil J. _The Search for Order: The Development of the Major Ideas in the Physical Sciences from the Earliest Times to the Present._ New York: Harper, 1960. Reprinted as _The Evolution of Physical Science_ New York: Grove Press, n.d. xvii + 398 pp.

An excellent textbook on the history of physics with some material on early astronomy and chemistry.

A104 Schonland, Basil. _The Atomists 1805-1933_. New York: Oxford University Press, 1958. x + 198 pp.

A105 Schroedinger, Erwin. _Science Theory and Man_. (Formerly published under the title _Science and the Human Temperament_, 1935) New York: Dover Publications, 1957. 223 pp.

Preface by E. Rutherford; Biographical Introduction by James Murphy. The 1957 reprint includes an additional essay on elementary particles.

A106 Schuster, Arthur. _The Progress of Physics during 33 years (1875-1908)_. Cambridge, Eng.: Cambridge University Press, 1911. New York: Arno Press, 1975. x + 164 pp.

A107 Segré, Emilio. _Personaggi e scoperte della fisica contemporanea_. Milan: Mondadori, 1976. 297pp. Translation: _From X-Rays to Quarks. Modern Physicists and their Discoveries_. San Francisco: W.H. Freeman & Co., 1980. xii + 337 pp.

Partly recollections and part personal historical reconstruction of events in modern physics. Emphasizes the author's personal experiencesin nuclear and elementary particle physics. Excellent popularization.

A108 Segré, Emilio. "La fisica dell'ultimotrentennio." _Problemi Attuali di Scienza e di Cultura_. Quaderno N. 228. Roma: Accademia Nazionale dei Lincei, 1977, pp. 9-19.

Surveys of development in physics in the last 30 years.

A109 Seidel, Robert W. _Physics Research in California: The rise of a leading sector in American Physics_. Ph.D. Dissertation, University of California at Berkeley, 1978.

For summary see _Dissertation Abstracts International_, 39 (1979): 5685-A.

A110 _Seminario di Storia della Fisica_. 1-2 giugno 1974, Università di Genova, Istituto di Scienze Fisiche, Gruppo di Storia della Fisica e seminario didattico dell'I.S.F. (1975). 199 pp.

A111 Shpol'skiĭ, E.V. *SorokLet Soovetskĭo Fiziki*. Moskva: Fizmatgiz, 1958. 87 pp.

Enlarged and completed version of an article on Soviet physics which appeared in *Uspekhi Fizicheskikh Nauk*, 63 (1957): 461-501.

A112 Shpol'skiĭ, E.V. "Stoletie Spektral'nogo Analiza. Iz Istorii Fiziki." *Uspekhi Fizicheskikh Nauk*, 69 (1959): 657-78.

A century of spectral analysis.

A113 Shpol'skiĭ, E.V. *Ocherki Po Istorii Razvitii͡a Sovetskoĭ Fiziki, 1917-1967*. Moskva: Nauka, 1969. 143 pp.

Essays on the historical development of Soviet physics.

A114 Siegbahn, Manne. "The state of physics in 1901." *Nobel: The Man and His Prizes*. Stockholm: Sohlmans Forlag, 1950, pp. 397-402.

A115 Silla, Lucio, ed. *Un secolo di progresso scientifico italiano, 1839-1939*. Società Italiana per il Progresso delle Scienze. Roma (1939). 7 vols.

Includes items A92 and A121.

A116 Singh, Virendra. "Towards a 'grand unification.'" *Science Today* (March 1980): 43-51.

Survey of theories of elementary particles and fundamental forces since 1896.

A117 Small, Henry. *Physics Citation Index 1920-1929*. Vol. I: Citation Index. Vol. II: Corporate Index, Source Index. Philadelphia: Institute for Scientific Information, 1981. Vol. I, 1931 cols. (5 per page); Vol. II, 142 cols. (5 per p.) + 1431 cols. (4 per p.).

An essential tool for research on the history of quantum theory and other areas of physics in the 1920s. Lists every work cited in one of the 16 major physics journals during the 1920s (including those published in earlier years), and allows one to locate all citations of each of these articles.

A118 Smith, Thomas M. "Physics." *Issues and Ideas in America*. Edited by Benjamin J. Taylor and T. J. White. Norman: University of Oklahoma Press, 1977, pp. 285-93.

A119 Smyth, H. D. "From x-rays to nuclear fission." *American Scientist*, 35 (1947): 485-501.

A120 Snow, C. P. *The Physicists*. Boston: Little, Brown, 1981. 192 pp.

A personal account of 20th century atomic physics, written "largely from memory"; copiously illustrated, published posthumously (a first draft was completed just before the author's death in 1980). This book was Snow's last heroic attempt to bridge the gap between his famous "two cultures" by explaining to the lay reader what it is like to be a scientist.

A121 Somigliana, Carlo, and Bruno Finzi. "Meccanica razionale e fisica matematica." *Un secolo di progresso* (item A115), pp. 211-242.

A122 *Sovetskaiâ Atomnaia Nauka i Tekhnika*. Edited by K. I. Shchelkin. Moskva: Atomizdat, 1967. 391 pp.

Soviet atomic science and technology.

A123 Spasskii, B. I. *Istorija fiziki...v cačestve ucebnogo posobija dlja*. Untov. I-II. Moskva, 1963. 331 + 300 pp.

A124 Taton, René, ed. *Histoire Générale des Sciences*. Tome III. *La Science Contemporaine*. Volume II, *Le XXe Siècle*. Paris: Presses Universitaires de France, 1964. Translation: *History of Science. Science in the Twentieth Century*. New York: Basic Books, 1966. xxiv + 638 pp.

Includes:L. de Broglie, "Contemporary Atomic and Quantum Physics"; Mme M.-A. Tonnelat, "Relativity"; A. Guinier, "Solid-State Physics"; V. Ronchi, "Optics"; J. P. Mathieu, "Spectroscopy"; G. Allard, "Thermodynamics"; E. Bauer and E. Herpin, "Magnetism"; P. Marzin and J. Le Mezec, "Electricity and Electronics"; J. Teillac, M. Langevin, P. Radvanyi, R. Nataf, L. Jauneau, "Radioactivity and Nuclear Physics"; P. Tardi, "Geophysics"; J. C. Pecker, "Theoretical Astrophysics"; P. Couderc, "Cosmology and Cosmogony."

A125 Taylor, J. G. *New Worlds in Physics*. London: Faber & Faber, 1974.

A popular survey, from Thales & Aristotle to quarks & quasars.

A126 Thomson, George. *Inspiraţie şi descoperire*. In româneşte de dr. C. Tănăsoiu şi R. Efraim. Bucureşti: Editura enciclopedită română, 1973. 205 pp. [Probably a translation of item B43]

A127 Toraldo di Francia, Giuliano. *L'indagine del mondo fisico*. Torino: Einaudi, 1976. xii + 606 pp.

* Trigg, G. L. Crucial experiments in modern physics. Cited herein as item JD20.

A128 Turkevich, John. "Soviet Physics and Chemistry." Soviet Science (item A22), pp.70-79.

A129 Ullmo, J. La pensee scientifique moderne. Reedition. Paris: Flammarion. 314 pp.

A130 Vinokurov, B. Z. "Nekotorye Voprosy Razvitiia Fiziki v Rossii v Nachale XX Veka." Voprosy Istorii Fiziki i ee Prepodavaniia. Tambov (1961), pp. 125-130.

On the development of physics in Russia at the beginning of the twentieth century.

A131 Vlachy', Jan. "Physics in Europe - sources of evidence." Czechoslovak Journal of Physics, sekce B, 25 (1975): 823-38.

A132 Vlachy', Jan. "Physics in Europe - more sources of evidence." Czechoslovak Journal of Physics, sekce B, 28 (1978): 928-47.

A133 Vlachy', Jan. "Publication output of Czechoslovak Physics." Czechoslovak Journal of Physics, sekce B, 28 (1978): 1409-12.

A134 Vucinich, Alexander. Science in Russian Culture, 1861-1917. Stanford: Stanford University Press, 1970. xv + 575 pp.

Chapter 12, "Modern Physics and Chemistry," pp. 362-96, includes remarks on A. F. Ioffe and P. Ehrenfest.

A135 Walker, Charles T., and Glen A.Slack. "Who named the -ON's?" American Journal of Physics, 38 (1970): 1380-89.

A136 Weart, Spencer R., ed. Selected Papers of Great American Physicists. New York: American Institute of Physics, 1976. 176 pp.

A137 Weart, Spencer R. "The last 50 years - a revolution?" Physics Today, 34, no. 11 (Nov. 1981): 37-49.

Survey of the history of physics since 1931.

A138 Webber, David, ed. Modern Physics: Selected Readings. Baltimore: Penguin Books, 1971. 319 pp.

Extracts from review articles and books published in the 1960s.

A139 Weber, R. L., and E. Mendoza, eds. A Random Walk in Science. New York: Crane, Russak & Co., 1973. xvii +

A139 Weber, R. L., and E. Mendoza, eds. A Random Walk in Science. New York: Crane, Russak & Co., 1973. xvii + 206 pp.

An anthology of humorous or odd writings, many by modern physicists.

A140 Weiner, Charles, ed. History of Twentieth Century Physics. New York: Academic Press, 1977. xii + 457 pp. (Proceedings of the International School of Physics "Enrico Fermi"; course 57)

A141 Weisskopf, Victor F. Physics in the Twentieth Century: Selected Essays. Cambridge, Mass.: MIT Press, 1972. xv + 368 pp.

Includes: Foreword by H. A. Bethe; "My Life as a Physicist"; "Niels Bohr, the Quantum, and the World"; "Marie Curie and modern science"; popular articles on recent developments in quantum theory, nuclear theory, elementary particles.

* Whitrow, Magda. Isis Cumulative Bibliography. Cited below as item B46.

A142 Whittaker, Edmund. A History of the Theories of Aether and Electricity. 2 vols. London: Nelson, 1951-53. Reprint: New York: Humanities Press, 1973. xiv +434 + xi + 319 pp.

Volume 2, "The Modern Theories 1900-1926" includes chapters on radioactivity, relativity, quantum theory, spectroscopy, gravity, matrix mechanics and wave mechanics.

A143 Whyte, Lancelot Law. "Pierre Curie's Principle of One-Way Process." Studium Generale, 23 (1970): 525-32.

A144 Wien, W. Vorträge über die neuere Entwicklung der Physik und ihrer Anwendungen. Leipzig: Barth, 1919. iv + 116 pp.

A145 Wiener, Philip P., ed. Dictionary of the History of Ideas. New York: Scribner, 1973. 4 vols.

Includes: M. Jammer, "Indeterminacy in physics," vol. II, pp. 586-94; B. Hoffmann, "Relativity," vol. IV, pp. 74-92; S. Bochner, "Space," vol. IV, pp. 295-307; M. Capek, "Time," vol. IV, pp. 389-98; G. J. Whitrow, "Time and Measurement," vol. IV, pp. 398-406.

A146 Wigner, E. "50 Years of principles of symmetry." (In Hungarian) Fizikai Szemle, 27, no. 8 (Aug. 1978): 281-287.

Brief review of the role of principles of symmetry before quantum mechanics noting the increased importance of symmetry principles in quantum mechanics in contrast to classical mechanics (in spectra, solid state, nuclear and particle physics).

A147 Yajima, Suketoshi. "Les sciences physiques au Japon durant l'ere de Meiji (1868-1912)." *Archives International d'Histoire des Sciences*, 9 (1956): 3-12.

A148 Zachoval, Ladislav. "O fyzikální vědecké práci mezi oběma světovými válkami v Praze." *Práce z déjin přírodnich věd*, 11 (1979): 67-78.

"On physics research in Prague between World Wars I and II.

A149 Ziggelaar, A. *Bronnen der Natuurkunde. Een verzameling historische artikeln*. Groningen: Wolters-Noordhoff N.V., 1971. 160 pp.

Anthology of 20 extracts from classic works in physics, from Archimedes to Mossbauer, with brief introductions; all in Dutch.

A150 Zirkle, Conway. "An Appraisal of Science in the USSR." *Soviet Science* (item A22), pp. 100-108.

B. BIOGRAPHIES
(Several subjects)

* Anderson, Carl D., et al. "Looking back on books and other guides." Cited below as item BA5.

B1 Barr, E. Scott. "Biographical material in the early issues of the *Physical Review*." *Bulletin of the American Physical Society*, series II, 11 (1966): 847.

B2 "Berliner Physiker in Bildern." *Physikalische Blaetter*, 23 (1967): 409-16.

B3 Bernstein, Jeremy. "A Scientific Education." *American Scholar*, 50 (1981): 237-43.

Recollections of Robert Oppenheimer, Julian Schwinger, Victor Weiskopf, and Philipp Frank as teachers.

B4 Bolton, Sarah K. *Famous Men of Science*. Third edition, revised by Edward W. Sanderson. New York: Crowell, 1946. 308 pp.

Includes chapters on the Curies, Karl and Arthur Compton, and Einstein.

B5 Broda, E. *The Intellectual Quadrangle: Mach - Boltzmann - Planck - Einstein*. Geneva: CERN Report 81-10 (July 1981). 20 pp.

Colloquium given at CERN in January 1981.

B6 Buckley, Paul, and F. David Peat, eds. *A Question of Physics: Conversations in Physics and Biology.* Toronto: University of Toronto Press, 1979. x + 159 pp.

Conversations with W. Heisenberg, L. Rosenfeld, P. A. M. Dirac, R. Penrose, J. A. Wheeler, C. F. von Weizsaecker, I. Prigogine, D. Bohm, etc.

B7 Cline, Barbara Lovett. *Men who made a New Physics: Physicists and the Quantum Theory*. (Originally published as *The Questioners*) New York: Crowell, 1965. Reprint, New York: New American Library, 1969. 223 pp.

Chapters on E. Rutherford, M. Planck, A. Einstein, N. Bohr, W. Pauli, W. Heisenberg.

B8 Ewald, Paul P. "Physicists I have known." *Physics Today*, 27, no. 9 (1974): 42-47.

B9 Fermi, Laura. *Illustrious Immigrants: The Intellectual Migration from Europe, 1930-41*. Second Edition. Chicago: University of Chicago Pres, 1971. xii + 431 pp.

Includes a chapter on the atomic scientists.

B10 Fleming, Donald, and Bernard Bailyn, eds. *The Intellectual Migration, Europe and America 1930-1960*. (Expanded edition of *Perspectives in American History*, 2) Cambridge, Mass.: Harvard University Press, 1969. 748 pp.

Includes: D. Fleming, "Emigre physicists and the Biological Revolution"; C. Weiner, "A New Site for the Seminar: The Refugees and American Physics in the Thirties"; biographical sketches of 300 emigres.

B11 Frisch, O. R., et al. *Trends in Atomic Physics: Essays dedicated to Lise Meitner, Otto Hahn, Max von Laue on the Occasion of their 80th Birthday*. New York: Interscience, 1959. 285 pp.

English translation of *Beiträge zur Physik und Chemie des 20. Jahrhunderts*.

B12 *Geist und Gestalt. Biographische Beiträge zur Geschichte der Bayerischen Akademie der Wissenschaften vornehmlich im zweiten Jahrhundert ihres Bestehens*. Muenchen: Beck, 1959. 318 + 297 + 269 pp.

Band II, "Naturwissenschaften," contains essays on physicists at the Bavarian Academy of Sciences, including Walther Gerlach on Boltzmann and Roentgen, Fritz Bopp on Sommerfeld, Walther Meisner on Carl von Linde, Oscar Knoblauch, Wilhelm Nusselt and Johann Ossanna; George Joos on Robert Emden; Eduard Ruechardt on W. Wien.

B13 Gillispie, Charles Coulston, ed. *Dictionary of Scientific Biography*. 16 vols. New York: Scribner's, 1970-80. xii + 624, x + 628, xiii + 624, xiv + 62, xiii + 624, xiii + 619, xii ++ 625, xii + 624, xiii + 620, xii + 622, xiii + 618, xiii + 620, xiii + 623, [xiii] + 640, xi + 818, [x] + 510 pp. (double-column pages)

The best single source for comprehensive, authoritative biographies of scientists. The symbol @ following a name in this bibliography indicates that an article may be found in the *Dictionary*.

B14 Haber, Louis. *Women Pioneers of Science*. New York: Harcourt Brace Jovanovich, 1979.

Juvenile book; includes chapters on Maria Goeppert Mayer and Rosalyn S. Yalow.

B15 Hartmann, Hans. _Schöpfer des neuen Weltbildes: Grosse Physiker unserer Zeit_. Bonn: Athenäum-Verlag, 1952; Hamburg & Berlin: Deutsche Hausbucherei, 1952. 352 pp.

Chapters on W. Roentgen, the Curies, Marconi, Planck, Einstein, Rutherford, M. v. Laue, O. Hahn, L. Meitner, Strassmann, Bohr, Heisenberg, de Broglie, F. Dessauer, W. Bothe, P. Jordan.

B16 Hermann, Armin. "Neue Quelle zur Quanten- und Relativitaetstheorie." _Physikalische Blaetter_, 23 (1967): 431-32; _Sudhoffs Archiv_, 51 (1967): 361-63.

Letters to A. Sommerfeld by physicists, 1913-46.

B17 Hermann, Armin. _German Nobel Prizewinners_. Munich: Heins Moos Verlagsgesellschaft, 1968. 172 pp.

B18 Hermann, Armin. "Physik als Philosophie und Weltgeschichte. Zu Leben und Werk von Einstein, Hahn, und Meitner." _Gedächtnisausstellung zum 100. Geburtstag ... Katalog_. Berlin, 1979, pp. 11-25.

Idem in _Physik und Didaktik_, 3(1980): 203-18.

B19 Hermann, Armin. "Mehr gemeinsam als nur die Zeit der Geburt. Zur Hundertjahrfeier fuer Albert Einstein, Otto Hahn, Lise Meitner, und Max v. Laue." _Mpg-Spiegel_, 6 (1978): 36-40.

Idem (slightly abbreviated) in _Umschau in Wissenschaft und Technik_, 1 (1979): 5 ff.

B20 Hermann, Armin. "Max Born, Niels Bohr, Werner Heisenberg. Die theoretische Physik im 20. Jahrhundert." _Die Grossen der Weltgeschichte_, Bd. XI. Zurich, 1978, pp. 35-49.

B21 Hermann, Armin. "Die Funktion von Briefen in der Entwicklung der Physik." _Berichte zur Wissenschaftsgeschichte_, 3 (1980): 55-64.

"Periodicals did not replace the letter, but only supplemented it. Any scholar who wants to participate in the discussion out of which science arises had to resort to the letter; the reading of journals did not suffice. This circumstance is illustrated in particular by the example of quantum theory and by Wolfgang Pauli."

B22 Jaffe, George. "Recollections of three Great Laboratories." _Journal of Chemical Education_, 29 (1952): 230-38.

The three are Ostwald's in Leipzig, J. J. Thomson's in Cambridge, and the Curies' in Paris. Includes also remarks about L. Boltzmann.

B23 Kaplan, Flora, comp. Nobel Prize Winners: Charts - Indexes - Sketches. Chicago: Nobelle Publishing Co., 1939. Second Edition, 1941. xvi + 144 pp.

The Physics section, pp. 13-38, is based on research by "a recent graduate in physics from the University of Chicago, who wishes to be identified as 'George Stoner.'"

B24 Kirsten, Christa, and Hans-Guenther Koerber, eds. Physiker ueber Physiker. Wahlvorschlaege zur Aufnahme von Physikern in die Berliner Akademie 1870 bis 1929, von Hermann v. Helmholtz bis Erwin Schroedinger. (Studien zur Geschichte der Akademie der Wissenschaften der DDR, Band 1) Berlin: Akademie-Verlag, 1975. 299 pp.

B25 Kirsten, Christa, and Hans-Guenther Koerber, eds. Physiker ueber Physiker II. Antrittsreden, Erwiderungen bei der Aufnahme von Physikern in die Berliner Akademie, Gedaechtnisrede 1870 bis 1929. (Studien zur Geschichte der Akademie der Wissenschaften der DDR, Band 8) Berlin: Akademie-Verlag, 1979. 298 pp.

Includes: R. Rompe & H.-J. Treder, "Die grosse Zeit der Physik in Berlin," 9-25; lectures by and about P. Drude, H. Rubens, K. Schwarzschild, A. Einstein, E. Schmidt, C. Caratheodory, M. von Laue, O. Hahn, F. Paschen, E. Schroedinger.

B26 Kleinert, Andreas. "Albert Einstein, Otto Hahn, Max von Laue, Lise Meitner, Zum 100 Geburtstag der Begrunder des Atomzeitalters." Buch und Bibliothek, 32 (1980): 507-17.

B27 Libby, Leona Marshall. The Uranium People. New York: Crane, Russak, 1979. x + 341 pp.

Described on the dust jacket as "The human story of the Manhattan Project by the woman who was the youngest member of the original scientific team." Includes recollections of L. Alvarez, H. L. Anderson, A. H. Compton, E. U. Condon, E. Fermi, K. Fuchs, Gen. L. R. Groves, D. Lilienthal, R. Mulliken, J. R. Oppenheimer, I. I. Rabi, E. Segre, L. Szilard, E. Teller, H. Urey, J. A. Wheeler, E. Wigner, W. Zinn and others.

B28 Livanova, A. Fiziki o Fizikakh. Moskva: Znamii͡a, 1964.

Includes sections on I. V. Kurchatov, A. F. Ioffe and L. I. Mandel'shtam based on recollections of colleagues and students (pp. 136-220).

B29 Livanova, A. "Fiziki o Fizikakh." _Puti v Neznaemoe_, Sb. 4. Moskva, 1964, pp. 417-54.

On encounters with I. V. Kurchatov, A. Einstein, L. I. Mandel'shtam.

B30 Ludovici, L. J., ed. _Nobel Prize Winners_. London: Arco, 1957; Westport, CT: Associated Publishers, 1957. xi + 226 p.

Includes articles by J. Bronowski on Einstein, and by N. Feather on Rutherford.

B31 MacCallum, T. W. and Stephen Taylor, eds. _The Nobel Prize-Winners and the Nobel Foundation 1901-1937_. Zurich: Central European Times Pub. Co., 1938. xi + 599 pp.

Biographies of physics prize-winners, 29-107; portraits and signatures of physics prize-winners, 391-436.

B32 Macorini, E., ed. _Scienziati e Tecnologi Contemporanei_. Volume I. Milano: Arnoldo Mondadori, 1974. 527 pp.

Includes brief biographies & autobiographies of H. Alfven, L. W. Alvarez, E. Amaldi, V. A. Ambartsumian, P. V. Auger, J. Bardeen, N. G. Basov, L. V. Berkner, G. Bernardini, H. A. Bethe, J. A. B. Bjerknes, P. M. S. Blackett, F. Bloch, D. I. Blokhintsev, N. N. Bogoliubov, N. H. Bohr, M. Born, W. Bothe, I. S. Bowen, W. L. Bragg, W. H. Brattain, P. W. Bridgman, L. V. de Broglie, E. C. Bullard, M. J. Buerger, V. Bush, H. B. G. Casimir, P. A. Cerenkov, J. Chadwick, S. Chapman, O. Chamberlain, S. Chandrasekhar, J. D. Cockcroft, A. H. Compton, E. U. Condon, L. Cooper, O. M. Corbino, R. H. Dalitz, C. G. Darwin, C. J. Davisson, P. J. Debye, M. Delbruck, R. H. Dicke, P. A. M. Dirac, A. S. Eddington, A. Einstein, L. Esaki, R. P. Feynman, G. N. Flerov, V. A. Fok, F. Franck, I. M. Frank, J. I. Frenkel, A. A. Fridman, O. R. Frisch, D. Gabor, G. Gamow, M. Gell-Mann, I. Giaver, W. F. Giauque, V. L. Ginzburg, D. A. Glaser, T. Gold, M. L. Goldberger, S. A. Goudsmit, O. Hahn, W. Heisenberg, G. L. Hertz, G. Herzberg, V. F. Hess, G. v. Hevesy.

B33 Macorini, E., ed. _Scienziati e tecnologi contemporanei_. Volume II. Milano: Arnoldo Mondadori, 1974. 527 pp.

Includes brief biographies and autobiographies of R. Hofstadter, F. Hoyle, E. P. Hubble, F. V. Hunt, L. Infeld, A. F. Ioffe, P. Jacquinot, J. H. Jeans, H. D. Jensen, F. Joliot-Curie, I. Joliot-Curie, P. Jordan, B. D. Josephspn, P. L. Kapitsa, T. von Karman, A. Kastler, A. Katzir-Katchalsky, K. I. Konstantinovich, V. N.

Kondrat'ev, H. A. Kramers, N. M. Krylov, I. V. Kurchatov, P. Kusch, W. E. Lamb, Jr., L. D. Landau, G. S. Landsberg, I. Langmuir, M. von Laue, E. O. Lawrence, P. P. Lazarev, T. D. Lee, G. Lemaitre, G. N. Lewis, I. M. Lifshitz, E. Majorana, L. I. Mandelshtam, M. G. Mayer, E. M. McMillan, L. Meitner, P. M. Morse, H. G. Moseley, R. L. Mossbauer, N. Mott, L. E. F. Neel, J. v. Neumann, G. Occialini, L. Onsager, J. R. Oppenheimer,, W. K. H. Panofsky, W. Pauli, L. C. Pauling, R. E. Peierls, G. Polvani, B. Pontecorvo, I. Prigogine, A. M. Prohorov, E. M. Purcell, I. I. Rabi, G. Racah, C. V. Raman, N. F. Ramsey, F. Rasetti, J. A. Ratcliffe, T. Regge, O. W. Richardson, L. Rosenfeld, B. B. Rossi, M. N. Saha, A. D. Sakharov, A. Salam, P. P. Savich, J. R. Schrieffer, E. Schroedinger, J. S. Schwinger, G. T. Seaborg, E. Segre, F. Seitz, W. Shockley, M. K. Siegbahn, F. Soddy.

B34 Macorini, E., ed. *Scienziati e tecnologi Contemporanei*. Volume III. Milano: Arnoldo Mondadori, 1974. 738 pp.

Includes brief biographies & autobiographies of O. Stern, F. Strassmann, L. Szilard, I. E. Tamm, E. Teller, G. P. Thompson, S. I. Tomonaga, C. H. Townes, G. E. Uhlenbeck, H. C. Urey, J. A. Van Allen, J. H. Van Vleck, S. I. Vavilov, E. T. S. Walton, A. M. Weinberg, V. F. Weisskopf, E. P. Wigner, C. N. Yang, H. Yukawa, E. K. Zavoiskii, F. Zernike.

B35 McClelland, David C. "On the psychodynamics of creative physical scientists." *Contemporary Approaches to Creative Thinking*. Edited by Howard E. Gruber, G. Terrell, and M. Wertheimer. New York: Atherton Press, 1962, pp. 141-74.

B36 Nachmansohn, David. *German-Jewish pioneers in science 1900-1933. Highlights in Atomic Physics, Chemistry, and Biochemistry*. New York: Springer-Verlag, 1979. xx + 338 pp.

Includes sections on James Franck, Max Born, and Gustav Hertz.

* Neu, John. *Isis Cumulative Bibliography*. Cited below as item B46.

B37 Recami, Erasmo. "I Curie, Hahn, e Fermi," *La Fisica nella Scuola*, 4 (1971): 17-21.

B38 Roe, Anne. "Analysis of group Rorschachs of physical scientists." Journal of Projective Techniques, 14 (1950): 385-98.

Discusses psychological tests on physicists and physical chemists at Berkeley, California Institute of Technology, Chicago, Columbia, Cornell, and Massachusetts Institute of Technology.

B39 Rosenfeld, Albert. "Giants of the Half Century: Science." Saturday Review/World, August 10, 1974, p. 85.

According to a survey of scientists, asked to name "the greatest scientific figure of the past half-century," Schrodinger and Fermi were ranked after Watson & Crick.

B40 Segre, E. G.; J. Kaplan, L. I. Schiff, and E. Teller. Great Men of Physics: The Humanistic Element in Scientific Work. Los Angeles: Tinnon-Brown, Inc., 1969. vi + 110 pp.

Includes chapter by Schiff on Newton, Einstein, and gravitation, and by Teller on Bohr & complementarity.

B41 Snow, C. P. Variety of Men. New York: Scribner, 1967. xii + 270 pp.

Includes chapters on Einstein and Rutherford.

* Snow, C. P. The Physicists. Cited above as item A120.

B42 Sokolvskaîa, Z. K. 300 Biografiĭ Uchenykh. Nauka, Moskva (1982). 389 pp.

Introducing the 300 biographies of scientists published in the period 1959-1980 in the series Nauchno-Biograficheskaîa Literatura sponsored by the Moscow Institute of History of Science and Technology of the USSR Academy of Sciences. Ch. 1 gives a history of the series. Ch. 2 contains the list of the biographies with complete bibliographical information (including price and number of copies), a brief index of each volume, translations and reviews.

Modern physicists whose biography was published in the series include: N. N. Andreev (1880-1970), N. Bohr, S. I. Vavilov, Van den Broek, Gibbs, L. V. Kirenskiĭ, T. P. Kravets, A.N. Krylov, V. L. Lëvshin, H. A. Lorentz, W. Ostwald, M. Planck, E. Rutherford, S. F. Rodionov, D. S. Rozhdestvenskiĭ, F. Soddy, Ia. I. Frenkel', C. P. Steinmetz, T. A. Edison, A. Einstein, B.N. Iur'ev, etc.
Ch. 3 deals with the perspectives of the series presenting the proposals of authors to the editors of the

series.
Ch. 4 gives a list of the members of the Editorial Board of the series from 1959 to 1980, and provides also a list of the authors with a biographical sketch and a list of the reviewers of the books of the series.

B43 Thomson, George. *The Inspiration of Science*. New York: Oxford University Press, 1961. Reprint, Garden City, N. Y.: Doubleday/Anchor, 1968. viii + 184 pp.

Includes biographical sketches of J. J. Thomson, F. W. Aston, R. A. Millikan, E. Rutherford, H. A. Lorentz, A. Einstein, Lord Rayleigh, C. T. R. Wilson, M. Planck, C. J. Davisson.

B44 Ulam, S. M. *Adventures of a Mathematician*. New York: Scribner's, 1976. xi + 317 pp.

Includes recollections of physicists at Los Alamos.

B45 Weber, Robert L. and J. M. A. Lenihan, eds. *Pioneers of Science - Nobel Prize Winners in Physics.* London: Institute of Physics & Adam Hilger, Ltd. (U. S. Distributor: Heyden & Son, Philadelphia), 1980. 280 pp.

B46 Whitrow, Magda, ed. *ISIS Cumulative Bibliography: A Bibliography of the History of Science Formed from ISIS Critical Bibliographies 1-90, 1913-1965*. London: Mansell, in conjunction with the History of Science Society, 1971-76. 3 volumes. *ISIS Cumulative Bibliography 1966-1975*. John Neu, editor. Volume 1. London: Mansell, in conjunction with the History of Science Society, 1980. Additional volumes in press.

BIOGRAPHIES AND COLLECTED WORKS

BA. Subjects: A

Abraham, H.

BA1 Henri Abraham, commémoration du centenaire de sa naissance, a l'École normale superieure, le 7 decembre 1968. Paris: École normale superieure, 1969. 55 pp.

Includes: Y. Rocard, "H. Abraham et le laboratoire de physique de l'École normale superieure"; A. Kastler, "H. Abraham et l'enseignement de la physique"; L. Brillouin, "Souvenirs sur H. Abraham"; B. Decaux, "H. Abraham et la mesure du temps" and other articles.

Allison, F.

BA2 Carr, Howard E. "Fred Allison." (obituary) Physics Today, 28, no. 1 (Jan. 1975): 107, 109.

Amaldi, E. HW

* Conversi, M., ed. Evolution of Particle Physics: A Volume Dedicated to Edoardo Amaldi in his 60th Birthday. Cited below as item JF15.

BA3 Schaerf, Carlo, ed. Perspectives of Fundamental Physics, dedicated to Edoardo Amaldi. New York: Harwood Academic Pubs., 1979. vii + 470 pp.

Includes: E. Segré, "Italian physics in Amaldi's time"; E. Amaldi, "The years of reconstruction."

BA4 Segré, E. "Per il settantesimo compleanno di Edoardo Amaldi." Giornale di Fisica, 20 (1979): 163-83.

Includes numerous photographs of Amaldi and other physicists.

Anderson, C. D. $36

BA5 Anderson, Carl D., Linus Pauling, Walter H. Brattain, Emilio Segré, Robert Hofstadter, Charles H. Townes, Hannes Alfvén, Ivar Giaever, Samuel C. C. Ting, Rosalyn Yalow, and Arno A.Penzias. "Looking back on books and other guides." Physics Today, 34, no. 11 (Nov. 1981), 247-61.

"Physics Today asked US scientists who have won Nobel Prizes in physics or work close to physics to comment on what books - or what discoveries, people or issues - of the last 50 years have most decisively influenced them..." (Editor's note preceding article)

Andreev, N. HW

BA6 Glekin, G. V. _Nikolai Nikolaievich Andreev. 1880-1970_. Seriia Nauchno-biograficheskaia Literatura. Moskva: Nauka, 1980. 87 pp.

Artsimovich, L. HW

BA7 _Vospominaniia ob akademike L. A. Artsimoviche_. Moskva: Nauka, 1981. 195 pp.

Recollections of Lev Andreevich Artsimovich (1909-1973) by A. I. Alikhan'ian, Hannes Alfven, A. P. Grinberg, F. A. Long, A. M. Petros'iants, Glenn T. Seaborg, Bernard T. Feld, Harold P. Furth with a biographical sketch and a collection of documents published by V. Ia. Frenkel' on L. A. Artsimovich's activity at the Leningrad Physico-Technical Institute.

BB. Subjects: B

Bethe, H. $67 HW

BB1 Bernstein, Jeremy. _Hans Bethe, Prophet of Energy_. New York: Basic Books, 1980. xii + 212 pp.

Based on articles published in _The New Yorker_, December 1979.

Bhabha, H. @ HW

* Anderson, Robert S. _Building Scientific Institutions in India: Saha and Bhabha_. Cited herein as item C3.

Bloch, F. $52

BB2 Chodorow, M., R. Hofstadter, H. E. Rorschach and A. L. Schawlow, eds. _Felix Bloch and Twentieth Century Physics_. Houston, TX: William Marsh Rice University, 1980. 247 pp.

BB3 Schiff, L. I., and R. Hofstadter. "Felix Bloch. A brief professional biography." _Physics Today_, 18, no. 12 (Dec. 1965): 42-43.

Blokhintsev, D.

BB4 Mayer, M. E. "Dmitrii Ivanovich Blokhintsev." _Physics Today_, 32, no. 7 (July 1979): 62-63.

Bogoliubov, N. HW

BB5 Zubarev, D. N. and B. V. Medvedev. "Novye Metody v Teoreticheskoĭ Fizike." Priroda, 9 (1958): 51-57.

On the works of N. N. Bogoliubov, who received the Lenin prize in 1958.

Bohr, N. @ $22 HW CB

BB6 Bohr, Niels. Collected works. General editor L. Rosenfeld. Vol. 1: Early work (1905-1911). Edited by J. Rud Nielsen. Vol. 2: Work on Atomic Physics (1912-1917). Edited by Ulrich Hoyer. Amsterdam: North-Holland, 1972, 1981. xviii + 608; xvi + 646 pp.

The first volume includes a biographical sketch and correspondence.

BB7 "Niels Bohr." Ceskoslovensky casopis pro fyziku, sekce A, 22 (1972): 511-18, 622-29.

BB8 Nil's Bor. Shizn' i Tvorchestvo. Collection of articles translated from Danish. B. G. Kuznetsov, ed. with commentaries by U. I. Frankfurt. Moskva: Nauka, 1967. 344 pp.

Includes: V. L. Ginzburg, "Pamîati Nil'sa Bora" [Recollection of Niels Bohr], 26-39; B. G. Kuznetsov, "Fundamental'nais fizicheskis Bora", 103-120; B. G. Kuznetsov, "Detstvo i Îunost Bora" [On Bohr's infancy and youth], 121-48.

BB9 Festskrift til Niels Bohr. Copenhagen: Pa Halvfjerdsarskeagen, 1955.

BB10 Belokon', V. A. "Nil's Bor v Gostîakh u Sovetskikh Uchenykh." Uspekhi Fizicheskikh Nauk, 76 (1962): 185-189.

On Niels Bohr's visit to the USSR, May 1961.

BB11 Kuznetsov B. G., ed. Razvitie Sovremennoi Fiziki. Moskva: Nauka, 1964. 330 pp. (AN SSSR, Institut Istorii Estestvoznaniîa I Tekhniki)

BB12 Physics Today, 16, no. 10 (October 1963): 21-64.

Articles by J. Rud Nielsen, F. Bloch, J. A. Wheeler, L. Rosenfeld, V. F. Weisskopf, in memory of Niels Bohr.

BB13 Rozental, S., ed. Niels Bohr: His Life and Work as Seen by his Friends and Colleagues. Amsterdam: North-Holland; New York: Wiley, 1967. 355 pp.

Articles by L. Rosenfeld, E. Rudinger, O. Klein, W. Heisenberg, H. B. G. Casimir, O. R. Frisch, A. Bohr, A. Pais, J. Kalckar, C. Moller, M. Pihl, V. F. Weisskopf, J. Pedersen, V. Kampmann, R. Courant, P. A. M. Dirac, H. H. Koch, W. Scharff, M. Andersen, H. Bohr.

BB14 Silverberg, Robert E. Niels Bohr: The Man who Mapped the Atom. Philadelphia: Macrae Smith Co., 1965. 189 pp.

"For grades 9 to 12."

Born, M. @ $54 HW

BB15 Born, Max. Mein Leben: Die Erinnerungen des Nobelpreistraegers. Muenchen: Nymphenburger Verlagshandlung, 1975. 399 pp. Translation: My Life: Recollections of a Nobel Laureate. New York: Scribner's, 1978. xi + 308 pp.

The author's English version has been edited by his son Gustav Born, who added a preface and a postscript about his father's last years.

BB16 Born, Max. Autobiografia di un fisico. Roma: Editori Riuniti, 1980. 435 pp.

BB17 Born, G. V. R. "Max Born: another impression." Science, 206 (1979): 636.

BB18 Hermann, Armin. "Max Born, Niels Bohr, Werner Heisenberg. Die theoretische Physik im 20 Jahrhundert." Berichte zur Wissenschaftsgeschichte, 1 (1978), 163-172.

BB19 Matthew, J. A. D. "Max Born, 1882-1970." Physics Education, 13 (1978), 251-54.

BB20 Mayer, Maria Goeppert. "Pioneer of Quantum Mechanics, Max Born, Dies in Göttingen." Physics Today, (March 1970): 97, 99.

Bose, J. C. @ HW

BB21 Lezhneva, O. A. "Dzhagadis Chandra Boze (K 100-Letiîv so Dnia Rozhdeniia)" Uspekhi Fizicheskikh Nauk, 67 (1959): 171-176.

Bose, S. N.

BB22 Blanpied, W. "Einstein as Guru? The case of Bose." Einstein: the first Hundred Years (item BE41), pp. 63-69.

BB23 Singh, Virendra, E. C. G. Sudarshan, Girijapati Bhattacharjee, N. D. Sengupta, and Parimalkanti Ghosh. "The eighty years of Satyen Bose." Science Today, (January 1974): 28-45.

BB24 Wilson, Mitchell. Passion to Know: The World's Scientists. Garden City, NY: Doubleday, 1972.

Interview with S. N. Bose, pp. 355-60.

Bozorth, R. M.

BB25 Wood, Elizabeth A. "Richard M. Bozorth." (Obituary) Physics Today, 34, no. 9 (Sept. 1981), 112, 114.

Bozorth (1896-1981) was known for his research in ferromagnetism.

Bragg, W. H. @ $15 HW

Bragg, W. L. @ $15 HW

BB26 Ewald, P. P. "Lawrence Bragg." (Obituary) Physics Today, 24, no. 10 (Oct. 1971): 64-67.

Braun, F. $09 HW

BB27 Kurylo, Friedrich, and Charles Susskind. Ferdinand Braun: A Life of the Nobel Prizewinner and Inventor of the Cathode-Ray Oscilloscope. Cambridge, Mass.: MIT Press, 1981. 304 pp.

Braun (1850-1919) shared the 1909 Nobel Prize with Marconi for his contributions to wireless telegraphy.

Breit, G.

BB28 Kihss, Peter. "Dr. Gregory Breit, early authority on atom weapons, is dead at 82." New York Times (22 Sept. 1981): A29.

Brillouin, L. HW

BB29 Thomas, L. H. "Leon Brillouin; theorist was also radio engineer." Physics Today, 23, no. 1 (Jan. 1970): 125, 127.

BB30 Kastler, A. "Zhizn' i tvorchestvo Leona Brilliuena." Uspekhi Fizichekikh Nauk, 106 (1972): 101-18.

Translated from L'Onde Electrique (1970).

Broglie, L. de $29 HW CB

BB31 Louis de Broglie: Sa Conception du Monde Physique; le Passe' et l'Avenir de la mecanique ondulatoire. Paris: Gauthier-Villars, 1973. xxviii + 387 pp.

Essays by M. A. Tonnelat, J. L. Destouches, L. de Broglie and others.

BB32 *Recherches d'un demi-siecle*. Paris: Michel, 1976. 411 pp.

Broglie, M. de @ HW CB

Brown, H.

BB33 Brown, Harold. "Confessions of an ex-physicist." *Physics Today*, 19, no. 6 (June 1966): 45-50.

On his experiences in government and as manager of large research projects.

Brown, S. C.

BB34 Allis, Wiliam P. "Sanborn Conner Brown" (Obituary) *Physics Today*, 35, no. 5 (May 1982): 98-99.

Brown, born in 1913, died 28 November 1981. In addition to his research in plasma physics and his international activities related to physics education, he wrote a major biography of Count Rumford and edited the reprint of Rumford's *Collected Works*.

Bruhat, G. CB

Burgers, J. M.

BB35 Koiter, W. T. "Johannes Martinus Burgers (13 januari 1895 - 7 juni 1981)" *Jaarboek van de Koninklijke Nederlandse Akademie van Wetenschappen*, 1981/82, pp. 1-6.

Burgers started his career with a contribution to the Rutherford-Bohr atomic model (1918), then turned to problems in the theory of turbulence and became a leader in fluid dynamics.

BC. Subjects: C

Chadwick, J. $35 HW

BC1 Feather, Norman. "Sir James Chadwick, Hon. F. R. S. E." (Obituary) *Yearbook of the Royal Society of Edinburgh*, (1975), pp. 10-12.

BC2 Oliphant, Mark. Obituary. *Physics Today*, 27, no. 10, (Oct. 1974): 87, 89.

Chapman, S. HW

BC3 Akasofu, Syun-Ichi, Benson Fogle, and Bernhard Haurwitz, eds. Sydney Chapman, Eighty, from his friends. Boulder, Colorado: National Center for Atmospheric Research and University of Colorado, 1968. xvi + 230 pp.

Articles about his work on kinetic theory of gases, geomagnetism and solar physics, ionospheric physics, etc.; personal recollections; list of publications.

Cherwell, Lord @ HW

BC4 Contemporary Scientific Archives Center, Oxford. "Collections completed, April-September 1981." Progress Report No. 16 (1981), pp. 3-4.

Reports that CSAC Catalogue No. 80/4/81 pp. 497, has been deposited in the Library of Nufield College, Oxford, and summarizes briefly the Cherwell Archive which is now open to researchers.

BC5 Thomson, George P. "Frederick Alexander Lindemann, Viscount Cherwell, 1886-1957." Biographical Memoirs of Fellows of the Royal Society, 4 (1958): 45-71.

Compton, A. @ $27 HW

BC6 Blackwood, J. R. The House on College Avenue. The Comptons at Wooster, 1891-1913. Cambridge: MIT Press, 1968. xxv + 265 pp.

On Karl T. Compton and Arthur Holly Compton.

BC7 Johnson, M., ed. The Cosmos of Arthur Holly Compton. New York: Knopf, 1967. 483 pp.

BC8 Shankland, Robert S., ed. Scientific Papers of Arthur Holly Compton. X-Ray and Other Studies. Chicago: University of Chicago Press, 1973. xxix + 777 pp.

Includes a sketch of Compton's career, 7 photographs, bibliography, and brief recollections by O. W. Richardson and A. Einstein.

Compton, K. @ HW

* Blackwood, J. R. The house on College Avenue. Cited herein as item BC6.

Condon, E. HW

BC9 Branscomb, L. M. (Obituary) Physics Today, 27, no. 6 (June 1974): 68-70.

* Cochrane, E. C. *Measures for Progress: A History of the National Bureau of Standards*. Cited herein as item C16.

BC10 Eisenhart, Churchill. "Edward Uhler Condon 1902-1974." *Dimensions* (National Bureau of Standards)(July 1974): 150-51, 166-67.

BC11 Wade, N. "Condon honored as early Nixon Victim." *Science*, 184 (1974): 544.

Cotton, A.

BC12 Rosmorduc, J. "Un grand physicien experimentateur: Aime Cotton (1869-1951)," *Fundamenta Scientiae*, 20 (1975): 1-19.

Cowan, C. CB

BC13 Reines, F. (Obituary) *Physics Today*, 27, no. 8 (August 1974): 68-69.

Curie, I.

BC14 Faĭnboĭm, I. B. *Iren I Frederik Zholio-Kiuri*. Moskva: Prosveshchenie,1964.64 pp. (Liudi Nauki I Tekhniki)

On Irène and F. Joliot-Curie.

Curie, M. S. @ $03 $11(Chemistry) HW CB

* Curie, Marie. *Pierre Curie*. Cited below as item BC30.

Includes 43 pages of "Autobiographical Notes" by Marie Curie.

BC15 Kiuri, Marie. *P'er Kiuri Marija o P'ere Kiuri. Iren u Frederic Žolio-Kiuri o Marii i P'ere Kiuri*. Perevod o francuscogo... Moskva: Nauka, 1968. 173 pp.

BC16 *Wklad Marii Sklodowskiej Curie do nauki. Skice monograficzne*. Warszawa: Panstwowe Wydawnictwo Naukowe, 1954. 270 pp.

BC17 Curie, Eve. *Madame Curie*. Paris: Gallimard, 1938. 374 pp. Reprint, 1981. Translation: *Madame Curie (1867-1934). A Biography*. Garden City, N. Y.: Doubleday, Doran, 1937. xv + 393 pp.

BC18 Giroud, Françoise. _Une femme honorable_. Paris: Fayard, 1981. 382 pp.

The author, a noted French journalist, retells the story of Madame Curie and Paul Langevin.

BC19 Ivimey, Alan. _Marie Curie, Pioneer of the Atomic Age_. New York: Praeger, 1969. vi + 122 pp.

BC20 Kolomy, Rudolf. "Marie Curie-Sklodowska." _Matematika a fyzika ve skole_, 5 (1974/1975): 57-61.

BC21 Latarjet,Raymond. "Rôle et influence de Marie Sklodowska-Curie dans les recherches biologiques." _Maria Sklodowska-Curie: Centenary Lectures_ (item BC28), pp. 125-135.

Survey on the development of Curie therapy and quantum radio-biology.

BC22 Matula, V. H. _Madame Curie_. Praha: Orbis, 1949. 32 pp.

BC23 Perrin, Francis. "L'Oeuvre scientifique de Marie Sklodowska-Curie et son influence sur les grande conquêtes de la physique moderne." _Maria Sklodowska-Curie: Centenary Lectures_ (item BC 28), pp. 13-23.

Deals with influence of M. Curie's work on the development of nuclear research in the decades following her fundamental researches.

BC24 Starosel'skaîa-Nikitina, O. A., and E. A. Starosel'skaia. "Obshchestvennye i Politicheskie Vzgliady Marii Sklodovskoi-Kiuri i ee Vklad v Mirovuiu i Natsional'nuiu Kul'turu Frantsii i Pol'shi." _Frantsuzskiĭ Ezhegodnik 1961_. Moskva, 1962, pp. 243-72.

On the social and political views of Marie Sklodowska-Curie and their impact on world culture, particularly French and Polish culture.

BC25 Zamfirescu, Eliza. _Marie Curie a nagy tudos etete es muve_. Bukarest: Forditas, 1957. 44 pp.

BC26 Zamfirescu, Eliza. _Viaţa şi opera savantei Marie Curie_. Bucureşti: Ed. tehnică, 1955. 48 pp. (Colecţia Societaţii pentru răspîndirea ştiinţei şi culturii)

BC27 Ziemecki, Stanislaw. "Sur l'oeuvre et la vie de Marie Sklodowska Curie." _Annales Universitatis Mariae Curie-Sklodowska_, Section AA, 22 (1967): 21-50.

See also other articles in this issue, devoted to celebrating the 100th anniversary of her birth.

BC28 _Maria Sklodowska-Curie: Centenary Lectures_. Vienna: International Atomic Energy Agency, 1968. 198 pp.

Proceedings of a symposium celebrating the centenary of the birth of Maria Sklodowska-Curie, held in Warsaw, 17-20 October 1967, and organized in Poland by the Maria Sklodowska-Curie Centenary Committee in co-operation with the International Atomic Energy Agency and the United Nations Educational, Scientific and Cultural Organization. Includes items BC21, BC23, JH40, JH41.

Curie, P. @ $03 HW CB

* Brouzenc, Paul. "Magnétisme et énergétique: La méthode de Duhem. A propos d' une lettre inedite de Pierre Curie." Cited below as item K22.

BC29 Bruzzaniti, Giuseppe. "'Storia reale' come ricostruzione di 'dizionari.' Un'ipotesi storiografica sull'impresa scientifica di Pierre Curie." _Scientia_, 115 (1980): 613-641. English translation, "'Real History' as 'Dictionary' reconstruction: A Historiographic Hypothesis for Pierre Curie's Scientific Undertaking, _ibid_: 643-61.

Discusses the unity of Pierre Curie's two research programs, (1) on symmetry of pyroelectric and magnetic properties of crystals; (2) on radioactivity.

BC30 Curie, Marie Sklodowska. _Pierre Curie_. With the auotbiographical notes of Marie Curie. Translated by Charlotte and Vernon Kellogg, with an introduction by Mrs. William Brown Meloney. New York: Macmillan, 1923. 242 pp. Reprint: Dover Publications, 1963.

Includes 43 pages of "Autobiographical Notes" by Marie Curie.

BC31 Grigor'yan, A. T., and O. A. Starosel'skaîa-Nikitina, eds._Trudy Instituta Istorii Estestvoznaniîa i Tekhniki_,

Vol. 19. Istoria Fiziko-Matematicheskikh Nauk. Moskva: Izd-vo AN SSSR, 1957. 723 pp.

Volume in honor of Pierre Curie. Papers by A. N. Nesmeianov, O. A. Starosel'skaîa-Nikitina, Ia. G. Dorfman, and I. I. Shafranovskii.

BD. Subjects: D

Dahl, O.

BD1 Festskrift til Odd Dahl. Bergen: A. S. John Griegs Boktrykkery, 1968. 103 pp.

Contains: Merle A. Tuve, "Odd Dahl at the Carnegie Institution, 1926-1936," 40-46; J. B. Adams, "Odd Dahl and the Machine at CERN," 83-92.

Darrow, K. K. HW

BD2 Van Vleck, J. H. "Karl Kelchner Darrow - Writer, Councilor, and Secretary." Physics Today, 20, no. 4 (April 1967): 23-26.

Debye, P. J. W. @ $36 HW CB

BD3 Corson, D. R., E. E. Salpeter, and S. H. Bauer. "Peter J. W. Debye: An Interview." Science, 145 (1964): 554-59.

BD4 Hund, Friedrich. "Nachrufe auf verstorbene Mitglieder: Peter Debye. Jahrbuch der Akademie der Wissenschaften in Göettingen (1966): 59-64.

BD5 Long, F. A. "Peter Debye--An appreciation." Science, 155 (1967): 979-80.

Dirac, P. A. M. $33 HW

BD6 Salam, Abdus, and E. P. Wigner, eds. Aspects of Quantum Theory. New York:Cambridge University Press,1972. xvi +268 pp.

"Dedicated to P. A. M. Dirac to commemorate his seventieth birthday and his contributions to quantum mechanics." Includes biographical articles by R. J. Eden and J. C. Polkinghorne, and J. H. Van Vleck; articles on quantum theory by J. Mehra, R. Jost, A. Pais, R. E. Peierls and others.

BD7 Salam, Abdus, et al. "The Banquet of the Symposium in Honour of Paul Dirac." The Physicist's Conception of Nature (item A80), pp. 805-19.

Brief remarks by C. P. Snow and others, about Dirac.

Donder, T. de

BD8 Tomassi, Witold. "Professor Teofil de Donder." Wiadomosci Chemie, 11 (1957): 481-86.

Drude, P. K. L. @

BD9 Falkenhagen, Hans. "Zum 100 Geburtstag von Paul Karl Ludwig Drude (1863-1906)." Forschungen und Fortschritte, 37 (1963): 220-221.

Dryden, H. L. HW

BD10 Smith, Richard K. The Hugh L. Dryden Papers, 1898-1965. A preliminary catalogue of the basic collection. Baltimore: Johns Hopkins University, Eisenhower Library, 1974. 165 pp.

Dunning, J. R.

BD11 "John R. Dunning, 69, dies; isolated Uranium isotope." New York Times (Aug. 28, 1975), 34.

Dyson, F.

BD12 Brower, Kenneth. The starship and the canoe. New York: Holt, Rinehart and Winston, 1978. 270 pp.

About Freeman Dyson and his son.

BE. Subjects: E

Eckart, C.

BE1 "Carl Eckart" (Obituary) Physics Today, 27, no. 1 (Jan. 1974): 87.

Brief announcement of his death on 24 October 1973 at age 71. See also ibid., 26, no.11 (Nov. 1973): 75, and 27, no. 5 (May 1974): 15.

Eddington, A. S.

BE2 Coleman, A. J. "More on Eddington." Physics Today, 34, no. 12 (Dec. 1981): 72.

Agrees with Nawrocki (item BE3) on physicists' response to A.S.E.

BE3 Nawrocki, Paul. "In praise of Eddington." Physics Today, 34, no. 12 (Dec. 1981): 72.

The scientific community should reconsider "its conviction of Eddington for insanity" in view of later vindication of some of his ideas.

Ehrenfest, P.

BE4 Frenkel, V. Ia. Paul Erenfest, second edition, revised and enlarged. Moskva: Atomizdat (1977). 192 pp.

BE5 Frenkel, V. Ia. "Paul Erenfest." Paul Erenfest, Otnositel nost' Kvanty Statistika. Edited by B. G. Kuznetsov and V. Ia.Frenkel. Moskva: Nauka, 1972, pp. 308-37.

Also contains V.Ia. Frenkel, "Lorents i Erenfest," ibidem, 227-32, and "Kommentarii," ibidem, 338-43.

BE6 Frenkel, V. Ia. "Predislovie redaktora." (Editor's Preface) Erenfest-Ioffe, Nauchnaia perepiska (Scientific Correspondence). Edited by V. Ia. Frenkel. Leningrad: Nauka, 1972, pp. 3-14.

V. Ia. Frenkel, "Kommentarii," ibidem.

BE7 Frenkel, V. Ia. Paul Erenfest. Moskva: Atomizdat, 1971. 144 pp.

First edition of Ehrenfest biography.

Einstein, A. $21 HW CB

BE8 Einstein, Albert. Aus meinen späten Jahren. Stuttgart: Deutsche Verlags-Anstalt, 1979. 275 pp.

Gives original German-language texts of essays previously published only in English. The earlier (German) edition of this book contained German retranslations from English. (Information from G. Holton)

BE9 Aichelburg, Peter C. and Roman U. Sexl, eds. Albert Einstein: His Influence on Physics, Philosophy and Politics. Braunschweig/Wiesbaden: Vieweg & Sohn, 1979. xv + 220 pp.

With contributions by Peter G. Bergmann, Hiroshi Ezawa, Walther Gerlach, Banesh Hoffmann, Gerald Holton, Bernulf Kanitschneider, Arthur I. Miller, Andre Mercier, Roger Penrose, Nathan Rosen, Dennis W. Sciama, Joseph Weber, Carl-Fredrich von Weizsaecker, John A. Wheeler and Wolfgang Yourgrau.

BE10 "Albert Einstein 1879-1955--Life in pictures." Physics Today, 32, no. 3 (1979): 25-33.

BE11 Amaldi, Edoardo, et al. Sette Lezioni su Einstein. Torino: Stampatori, 1980. 190 pp.

Includes: T. Regge, "La relativita generale nell - astrofisica e nella cosmologia"; V. de Alfaro, "Relativita e fisica delle particelle fondamentali"; M. Rasetti, "Il contributo di Einstein alla meccanica statistica"; F. Pacini, "Astronomia relativistica dai quasar ai buchi neri"; E. Amaldi, "Onde gravitazionali"; L. Geymonat, "Il significato filosofico del pensiero di Einstein"; G. Dionigi, "I rapporti di Einstein con la scuola matematica italiana: Ricci-Curbastro, Bianchi, Levi-Civita"; reproduction in facsimile of a few letters between Einstein and Levi-Civita.

BE12 Angoff, Charles, ed. Science and the human imagination: Albert Einstein. Rutherford, N. J.: Fairleigh Dickinson University Press, 1978. 94 pp.

Includes lectures by J. Bernstein and G. Feinberg.

BE13 Barker, Peter. Albert Einstein's philosophy of science: A preliminary survey. Ph. D. Dissertation, State University of New York, Buffalo, 1975. 166 pp.

BE14 Bičák, Jiří. Einstein a Praha K 100. výročí narození Alberta Einsteina. Prague: Jednota Československých Matematiků a Fyziků, 1979. 63 pp.

BE15 Born, Max. "Erinnerungen an Einstein." Physikalische Blaetter, 21 (1965): 299-306.

BE16 Born, Max, ed. Albert Einstein -- Hedwig und Max Born: Briefwechsel 1916-1955. Munich: Nymphenburger, 1969. 330 pp. Translations: The Born-Einstein Letters. New York: Walker, 1971. x + 240 pp. Scienza e Vita, Lettere 1916-1955. Torino: Einaudi, 1973. 279 pp.

"If anyone needs a proof that science is a fundamental part of any serious unified culture in this century, here it is. This is the correspondence ... between two of the noblest men of the age. ... All that they said to each other should be the concern of educated men everywhere." -- C. P. Snow, review in Financial Times, reprinted in Washington Post, 14 Aug. 1971, p. E4.

BE17 Born, Max. "In memory of Einstein." Interdisciplinary Science Review, 3 (1978): 267-74.

BE18 Brdicka, Miroslav. "Einstein a Praha." Ceskoslovenský časopis pro fyziku, sekce A, 29 (1979): 269-75.

BE19 Broad, Wiliam J. "On centennial fever and its ironies." *Science*, 203 (1979): 1225.

* Byrne, Patrick H. "Statistical and causal concepts in Einstein's early thought." Cited below as item DB11.

BE20 Byrne, Patrick H. *Einstein's quest for the foundations of science*. Ph. D. Dissertation, State University of New York at Stony Brook, 1978. 331 pp.

BE21 Byrne, Patrick H. "The origins of Einstein's use of formal asymmetries." *Annals of Science*, 38 (1981), 191-206.

BE22 Caldirola, Piero. "L'opera scientifica di Albert Einstein." *Albert Einstein nel 25 anniversario della morte*. Milano: Istituto Lombardo di Scienze e Lettere, 1981, pp. 33-54.

BE23 Calinger, Ronald. "Albert Einstein: Theoretical physicist and humanitarian." *Methodology and Science*, 12 (1979), pp. 109-64.

BE24 Clark, Ronald W. *Einstein: The Life and Times*. New York: World, 1971. xv + 718 pp. Translation: *La Vita Pubblica e Privata del più grande scienziato del nostro tempo*. Milano: Rizzoli, 1976. 743 pp.

An excellent, comprehensive account for the layman.

BE25 Dirac, P. A. M., et al. "The Einstein century: Four generations of revolutionary thought." *Impact of Science on Society*, 29 (1979): 3-93.

BE26 Durrenmatt, F. *Albert Einstein*. Paris: Lettres Universelles, Ed. de l'Aire, Diff. PUF, 1982. 70 pp

Fantasy which includes Spinoza, Hume and Kant as protagonists.

BE27 *Einshtein i Razvitie Fiziko-matematicheskoi mysl; Sbornik Statei*. Moskva: Izd-vo Akademii Nauk SSSR, 1962. 237 pp.

BE28 *Einshtein i sovremennaia fizika. Sbornik nomiati A. Einsteina*. Moskva: Gostekhizdat, 1956. 260 pp.

BE29 Elkana, Yehuda, and Adi Ophir. *Einstein 1879-1979*. Exhibition, Jewish National and University Library. Jerusalem: Hebrew University, 1979. Text in English and Hebrew.

BE30 Erenburg, I. G. _Liûdi, Gody, Zhizn'_. Kn. 5 i 6, Moskva (1966), pp. 371-380.

Pages with recollections of Einstein.

BE31 Feuer, Lewis S. _Einstein and the Generations of Science_. Second edition. New Brunswick, NJ: Transaction Books, 1982. 390 pp.

BE32 Finis, Francesco de, ed. _Albert Einstein, 1879-1979: Relativity, quanta, and cosmology in the development of the scientific thought of Albert Einstein._ New York: Johnson Reprint, 1979. 825 pp.

Includes articles by E. Amaldi, G. Pizella, L. De Broglie, A. Kastler, and several others on current status of physical theories.

BE33 Forman, Paul. "Einstein and Newton: Two Legacies." _Wilson Quarterly_, Winter 1979: 107-14.

Includes Einstein's 1927 article on Newton.

BE34 Frank, Phillipp. _Einstein, his life and times_. Translated from a German manuscript by George Rosen. Edited and revised by Shuichi Kasaka, with a New Introduction. New York: Knopf, 1947, 1967. xxvii + 298 + xii pp.

An important work by the distinguished philosopher of science who knew Einstein throughout his career.

BE35 Frenkel, V. Ia., Iavelov, B. E. _Eĭnshteĭn - Izobretatel_. Moskva: Nauka, 1981. 160 pp.

Einstein the experimenter. Fluctuations and potential voltage-multiplier; Ampere's molecular currents. Heat pumps and absorption refrigerators.

BE36 Frenkel, V. Ia, and N. Ia. Moskovchenko. "Neizvestnoe pis'mo A. Einshteina." _Voprosy istorii estestvoznaniîa i tekhniki_, 2 (1981): 142.

Publication of a letter of 23rd February 1925 from Einstein to Ioffe with a recommendation for Ia. P. Grommer. Followed by V. Ia. Frenkel, "Kommentarii" (Comments), pp. 142-3.

BE37 Frenkel, V. Ia. and B. E. Iavelov. "Izobretatel'skaîa deîatel'nost' A. Eĭnshteĭna." _Einshteinovskii Sbornik 1977_, pp. 214-256.

Technical interests of Einstein in Bern.

BE38 Frenkel, V. Ia. "Einstein i sovetskie fiziki." Voprosy istorii estestvoznaniia i tekhniki, 3, no. 52 (1976): 25-30.

Fragments of correspondence of Einstein with Soviet physicists and recollections of him.

BE39 Ganev, I. Kh. "Vospominaniia o Professore Al'berte Einshteine." (translated from Bulgarian) Einshtein I Razvitie Fiziki. G. M. Idlis, ed. Moskva, (1981).

Recollections of Einstein.

BE40 Gingerich, Owen. "Albert Einstein: A Laboratory in the Mind." Science Year 1980 (pub. 1979), 395-410.

BE41 Goldsmith, Maurice, Alan McKay, and James Woudhuysen, eds. Einstein: The First Hundred Years. New York: Pergamon Press, 1980. xiii + 200 pp.

Articles by C. P. Snow, R. Furth, L. Infeld, P. A. M. Dirac, D. Bohm & B. Hiley, W. Blanpied, J. Wheeler, and others.

BE42 Goodstein, Judith. "Albert Einstein in California." Engineering and Science (California Institute of Technology), May-June 1979, pp. 17-19

BE43 Havranek, J., M. Solc, and J. Grygar. "V Praze o Einsteinovi a o Einsteinovi v Praze." Vesmir, 58 (1979): 178-83.

Relations of Einstein with Prague.

BE44 Hermann, Armin, ed. Albert Einstein/Arnold Sommerfeld. Briefwechsel. 60 Briefe aus dem goldenen Zeitalter der Modernen Physik. Basel (1968). 126 pp. There is also a Japanese edition(Tokyo, 1970).

BE45 Hermann, A. "Einstein und die deutschen Physiker." Naturwissenschaftliche Rundschau, 22 (1969): 109-111; also in Tribune. Zeitschrift zum Verstandnis des deutschen Judentums, 8 (1969): 3180-3183.

BE46 Hermann, Armin. "Einstein in Berlin." Jahrbuch Preussischer Kulturbesitz, 8 (1970): 90-114.

BE47 Hermann, Armin. "Albert Einstein. Umsturz im Weltbild der Physik." Die Grossen der Weltgeschichte, Bd. XI, Zurich (1978), 15-33.

BE48 Hermann, Armin. Die Neue Physik. Der Weg in das Atomzeitalter. Zum Gedenken an Albert Einstein. Max von Laue. Otto Hahn. Lise Meitner. Bonn: Inter Nationes,

1979. 170 pp. Translation: The new physics: the route into the atomic age: in memory of Albert Einstein, Max von Laue, Otto Hahn, Lise Meitner. Bonn: Bad Godesberg: Inter Nationes, 1979. 170 pp.

BE49 Hermann, Armin. "Eine neue Grösse der Weltgeschichte: Albert Einstein." Bild der Wissenschaft, 3 (1979): 59-71.

BE50 Hermann, Armin. "Einstein und kein Ende." Physikalische Blätter, 38, nr. 2 (Feb. 1982): 36-41.

BE51 Herneck, F. Albert Einshtein. Zizn vo imia istiny, gumanizma i spravedlivost. Moskva: Progress, 1966. 244 pp.

BE52 Holton, Gerald. "Einstein's search for the Weltbild." Proceedings of the American Philosophical Society, 125 (1981): 1-15.

BE53 Holton, Gerald, and Yehuda Elkana, eds. Albert Einstein: Historical and Cultural Perspectives. Princeton, NJ: Princeton University Press, 1982. xxxii + 439 pp.

Proceedings of the Jerusalem Einstein Centennial Symposium, 14-23 March 1979. Includes papers by G. Holton, A. I. Miller, P. G. Bergmann, M. J. Klein, M. Jammer, P. A. M. Dirac, B. Hoffmann, L. R. Graham, R. Jakobson, E. H. Erikson, I. Berlin, and others.

BE54 Illy, Jozsef. "Albert Einstein a Praha." Dějiny věd a techniki, Praha, 12 (1979): 65-80.

This is a full text of the paper "Albert Einstein in Prague," Isis, 70(1979), 76-84, without the chapters on Einstein's scientific activity in Prague (information from author).

BE55 Illy, Joseph. "Einstein és Spinoza." (Einstein and Spinoza) Magyar Filozófiai Szemle, 10 (1966): 159-165.

BE56 Ivanenko, D. D. et al. "K 100-Letiiu so dnia roshdeniia A. Einshteina." Voprosy Istorii Estestvoznaniia i Tekhniki, 3(67)-4(68) (1980): 3-71.

On the 100th anniversary of the birth of A. Einstein -- articles by D. D. Ivanenko, A. Trautman, V. P. Vizgin, D. Schulze, M. A. El'iashevich, B. E. Iavelov, I. S. Alekseev, G. E. Gorelik, G. E. Grigor'ev; unpublished letter of Einstein to I. Obreimov (1929).

BE57 Kedrov, B. M., Ovchinnikov, N. F., eds. _Printsip Simmetrii Nauka_. Moskva (1978). 397 pp.

Collection of papers dealing with Einstein, by I. S. Alekseev; I. A. Akchurin; Iu. I. Kulakov; V. M. Shemiâkinskiĭ; N. P. Konopleva; V. P. Vizgin; A. Tursunov and U. Radzhabov; N. F. Ovchinnikov.

BE58 Kinnon, Colette, A. N. Kholodin, and J. G. Richardson, eds. _The Impact of Modern Scientific Ideas on Society. In Commemoration of Einstein._ (Papers presented at the UNESCO Symposia, Munich-Ulm, September 1978,and Paris, May 1979) Boston: Reidel, 1981. xiv + 203 pp.

Includes lectures by P. A. M. Dirac, A. Salam, P. L. Kapitza, J. Ehlers, E. N. Hiebert, O. Pedersen, T. F. Nonnenmacher, C. W. Misner, and others.

BE59 Kobzarev, I. Iu. "Eĭnshteĭn i teoreticheskaiâ Fizika pervoĭ treti XX veka." _Eĭnshteĭn i Sovremennaiâ Fizika_, Moskva, 1979, pp. 3-21.

Einstein and theoretical physics in the first third of the 20th century.

BE60 Kolata, G. B. "Einstein skeptical of ESP after all." _Science_, 197 (1977): 1349.

BE61 Kuznetsov, B. G. _Albert Ainshtain_. Sofia: Tekhnika, 1964. 423 pp. Translations: _Einstein_, Moscow: Progress Publishers, 1965, 380 pp.; _Albert Einstein_, Warszawa: Panstw. wyd-wo naukowe, 1966, 517 pp.; _Einstein, Sa Vie, sa Pensee, ses Theories_, Paris: Verviers, Yerard, 1967, 344 pp.; _Albert Einstein_, Bucuresti: Editura Siintifica, 168, 492 pp. Fifth Russian edition, revised and enlarged, _Einshtein: Zhizn', Smert', Besmertie_, Moscow, 1979. 680 pp.

BE62 Kvasnica, Jozef. "Nerelativisticke dilo Alberta Einsteina." _Československy časopis pro fyziku_, sekce A, 29 (1979), 212-23.

Non-relativistic work of A. E.

BE63 Lambrecht, Jena H. "Zum Umschlagbild: Albert Einstein." _Die Sterne_, 51, Heft 2 (1975), 65-68.

BE64 Levinger, Elma Ehrlich. _Albert Einstein_. Milano: Mondadori, 1951. 169 pp.

BE65 Lvov, Vl. *Viaţa lui Albert Einstein*. Translated by M. Mayer & R. Weiner from the second Russian edition. Bucureşti: Ed. Ştiinţifică, 1960. 317 pp.

BE66 Nathan, Otto. "The Einstein Papers." *Science*, 213 (1981): 1313-1314.

BE67 Ne'eman, Yuval, ed. *To Fulfill a Vision*. Jerusalem Einstein Centennial Symposium on Gauge Theories and Unification of Physical Forces. Reading, MA: Addison-Wesley, 1980. xxxi + 279 pp.

Includes historical articles reprinted from journals, and papers by C. N. Yang, P. G. Bergmann and F. Gursey which allude to Einstein's work.

BE68 Nelkowski, H., A. Hermann, H. Poser, R. Schrader and R. Seiler, eds. *Einstein Symposium, Berlin*. New York: Springer, 1981. 550 pp.

BE69 Nisio, Sigeko. "The transmission of Einstein's work to Japan." *Japanese Studies in History of Science*, 18 (1979): 1-8.

On Ayao Kuwaki (1878-1945) and Jun Ishiwara (1881-1947).

BE70 Okamoto, Ippei. "Albert Einstein in Japan: 1922." Translated from the Japanese by Kenkichiro Koizumi. *American Journal of Physics*, 49 (1981): 930-40.

"Okamoto, a painter in the Western tradition who was employed at the time as a cartoonist for the Asahi newspaper in Toyko, traveled with the Einstein party" and recorded his impressions in words and sketches.

BE71 Pais, A. *'Subtle is the Lord ...' The Science and Life of Albert Einstein*. New York: Oxford University Press, 1982. xvi + 552 pp.

BE72 Pais, Abraham. "How Einstein got the Nobel Prize." American Scientist, 70 (1982): 358-65.

"Why did the Nobel Committee for Physics wait so long before giving Einstein the Prize and why did they not award it for relativity? ... It is understandable that the Academy was in no hurry to award the Prize for relativity until the experimental issues had been clarified. It was also the Academy's bad fortune not to have among its members in those early days anyone who could competently evaluate the content of relativity theory. Oseen's proposal to give Einstein the award for the photoelectric effect must have come as a relief of conflicting pressures."

BE73 Pantaleo, Mario, ed. Centenario di Einstein 1879-1979. Astrofisica e Cosmologia, Gravitazione, Quanti e Relativita negli sviluppi del pensiero scientifico di Albert Einstein. Firenze: Giunti Barbera,1979. 1181 pp.

Essays by A. Carrelli, E. Amaldi, P. G. Bergmann, B. Bertotti, H. Bondi, P. Caldirola, L. V. de Broglie, L. D. Faddeev, J. Isenberg, D. D. Ivanenko, P. Jordan, A. Kastler, A. Lichnerowiz, Ch. Møller, J. V. Narlikar, G. Pizzella, L. Radicati di Brozolo, E. Recami, R. Ruffini, H. Sato, D. W. Sciama, I. I. Shapiro, M. M. Shapiro, R. Silberberg, H. J. Treder, J. A. Wheeler.

BE74 Papp, D. Einstein, Historia de un espiritu. Madrid: Espassa-Calpe, 1979. 277 pp.

BE75 Perlmutter, A., and L. F. Scott, eds. On the Path of Albert Einstein. (Proceedings of a Congress, Coral Gables, Fla., January 1979). New York: Plenum, 1979. 180 pp.

BE76 Poss, Ondrej. "Albert Einstein." Elektron, nr. 3 (1979): 36-38.

BE77 Pyenson, Lewis. "L'education scientifique du jeune Einstein." Spectre, 8, no. 3 (1979): 5-14.

BE78 Quasha, Solomon. Albert Einstein: an intimate portrait. Larchmont, NY: Forest Pub. Co., 1980. xi + 332 pp.

BE79 Rassias, Themistocles M., and George M. Rassias, eds. Selected Studies: Physics-Astrophysics, Mathematics, History of Science, A Volume dedicated to the Memory of Albert Einstein. Amsterdam: North-Holland Pub. Co., 1982. xii + 392 pp.

Includes: S. Bourne, "Four conversations with Albert Einstein"; L. Debnath, "Albert Einstein -- Scientific Epistemology"; J. Dieudonne, "Les Grandes Lignes de l'Evolution des Mathematiques"; L. Pyenson, "Mathematics,

Education, and the Goettingen Approach to Physical Reality, 1890-1914"; E. Teller, "Why Einstein should have won the Nobel Prize"; S. Weinberg, "Einstein and Spacetime: Then and Now."

BE80 Regge, Tullio. "Albert Einstein nel centenario della nascita." *Rendiconti Accademia Nazionale delle Scienze detta dei XL, Memorie di Scienze Fisiche e Naturali 98º (1979-80)*, Vol. IV, fasc. 5: 41-48.

BE81 Resnick, Robert. "Misconceptions about Einstein: His work and Views." *Journal of Chemical Education*, 57 (1980): 854-62.

BE82 Rollnik, H. "Zum Artikel von A. Unsöld, Albert Einstein--Ein Jahr danach." *Physikalische Blätter*, 37, nr. 3 (1981): 65.

The President of the German Physical Society "distances himself" from the attack on Einstein by Unsold, cited below as item BE96.

BE83 Rumer, Iu. B. "Krupneĭshiĭ Fizik Sovremennosti." *Sibirskie Ogni*, 6 (1955): 156-164.

On Einstein.

BE84 Schlicker, Wolfgang. "Albert Einstein a politicka reakce v Nemecku po prvni svetove valce." *Pokroky matematiky, fyziky a astronomie*, 24 (1979): 153-60.

A. E. and political reaction in Germany after World War I.

BE85 Seidelmann, P. K., R. E. Schmidt and R. Berks. "Putting the stars at Einstein's feet." *Sky and Telesope*, 59, no. 3 (March 1980): 203-6.

On the memorial in Washington, D.C.

BE86 Smith, Richard D. "Hollywood's neglected genius." *Playboy*, 20, no. 6 (June 1973): 147-148, 150, 179.

BE87 Stachel, John. "Exploring the Man Beyond the Myth: Albert Einstein." *Bostonia Magazine*, 5, no. 1 (Feb. 1982): 8-17.

BE88 Stockton, William. "Celebrating Einstein." *New York Times Magazine*, 18 (February 1979): 16.

A sour survey of the centennial celebrations.

BE89 Sullivan, Walter. "An editor for Einstein's papers chosen after six-year search." *New York Times*, 5 (July 1976): 32.

Selection of J. Stachel.

BE90 Swenson, Loyd S., Jr., C. P. Snow, Howard Stein and Ilya Prigogine. "Albert Einstein: Four Commemorative Lectures." *Library Chronicle of the University of Texas at Austin*, New Series, Number 12 (1979): 31-91.

BE91 Tamm, I. E., G. I. Naan, and U. I. Frankfurt, eds.*Eĭnshteĭnovskiĭ Sbornik, 1967*. Moskva: Nauka, 1967. 370 pp.

Contains translations of Einstein-Solovine, and Einstein-Hadamard correspondence, and articles by V. Ginzburg and B. G. Kuznetsov.

BE92 Tamm, I. E., G. I. Naan, and U. I. Frankfurt, eds. *Eĭnshteĭnovskiĭ Sbornik, 1968*. Moskva: Nauka, 1968. 287 pp.

BE93 Tamm, I. E., G. I. Naan and U. I. Frankfurt, eds.*Eĭnshteĭnovskiĭ Sbornik, 1969-70*. Moskva: Nauka, 1970. 408 pp.

BE94 Tobey, Ronald C. *The American Ideology of National Science, 1919-1930.*. Pittsburgh: University of Pittsburgh Press, 1971. xiii + 263 p.

Chapter 4: "The Einstein controversy 1919-1924."

BE95 Úlehla, Ivan. "K stému výročí narození A. Einsteina." *Nová mysl*, 32 (1979): 125-28.

BE96 Unsöld, Albrecht. "Albert Einstein--Ein Jahr danach." *Physikalisch Blätter*, 36, nr. 11 (1980): 337-39.

An attack on Einstein which caused a heated controversy in Germany; see the news item in *Nature*, 290 (1981), 535, the announcement by Rollnik (item BE82) and the article by Hermann (item BE50.)

BE97 Valcovici, Victor. *Albert Einstein, teoriile sale, viaţa sa*. Bucureşti,1957. 35 pp. (Colecţia Societăţii pentru răspîndirea ştiinţei şi culturii, 248).

BE98 Vallentin, Antonina. *Le drame d'Albert Einstein*. Paris: Plon, 1954. 312 pp.

BE99 Wade, N. "Brain that rocked physics rests in cider box." *Science*, 201 (1978): 696.

BE100 Warner, Jack, Jr. "Dr. Einstein's Marvelous Flying Machine." Reader's Digest, March 1982, pp. 65-67. Condensed from "Westwinds," Westways, 73, no. 11 (Nov. 1981): 20, 78-79.

On Einstein's visit to Warner Brothers Studios in 1931.

BE101 Winters, Shelley. "Shelley Winters tells all about Sinatra, Brando, Monroe, Gable, Flynn..." Ladies Home Journal, 97, no. 5 (May 1980): 42-47, 156-68.

On Marilyn Monroe's fantasy about having an affair with Einstein. (see p. 166)

BE102 Woolf, Harry, ed. Some Strangeness in the Proportion. A Centennial Symposium to celebrate the achievements of Albert Einstein. Reading, Mass.: Addison-Wesley, 1980. xxxi + 539 pp.

Major historical papers by F. Gilbert, H. Woolf, G. Holton, A. I. Miller, M. J. Klein, A. Pais, S. S. Chern; many other papers on current problems growing out of Einstein's work, short comments and discussion.

BE103 Yang, Chen-Ning. "Einstein's impact on theoretical physics." Physics Today, 33, no. 6 (June 1980): 42-49.

BE104 "Z Einsteinovych uvah a dopiso." Ceskoslovensky casopi pro fyziku, sekce A, 29 (1979): 275-85.

"From deliberations and letters of A. E."

BE105 Zaikov, R. G. "Vospominaniia Ob Al'berte Einshteine." (translated from Bulgarian) Voprosy Istorii Estestvoznaniia i Tekhniki, 18 (1965): 16-19.

BE106 Zajac, Rudolf. "Albert Einstein a fyzika 20. storocia." Pokroky matematiky, fyziky a astronomie, 24 (1979): 61-76.

BE107 Zweifel, P. F. "The scientific work of Albert Einstein." Annals of Nuclear Energy, 7 (1980): 79-87.

Enskog, D.

BE108 Chapman, S. "Prof. David Enskog." Nature, 161 (1948): 193-94.

BE109 Faxen, Hilding. "Enskog, David." Svenskt Biografiskt Lexikon. Vol. 13. Stockholm: Bonnier, 1950, pp. 765-67

An English translation is available as a U.S. Government document, UCRL-6316.

Everett, H., III

BE110 "Dr. Hugh Everett III, Founder of Data Firm." Washington Post, 23 July 1982: B5.

Obituary of a co-founder of the "many-worlds" interpretation of quantum mechanics.

Exner, F. S.

BE111 Karlik, Berta, and Erich Schmid. Franz Serafin Exner und sein Kreis. Ein Beitrag zur Geschichte der Physik in Oesterreich. Wien: Verlag der Oesterreichischen Akademie der Wissenschaften, 1982. 168 pp.

Includes information on the Instituts fuer Radiumforschung, S. Meyer, E. v. Schweidler, E. Haschek, H. Mache, F. v. Lerch, V. Hess, F. Kohlrausch, E. Schroedinger, M. v. Smoluchowski, F. Hasenoehrl and others.

BF. Subjects: F

Fermi, E @ $38 HW CB

BF1 Amaldi, Edoardo. "Enrico Fermi." La Ricerca Scientifica, 1 (1955): 5-15.

Obituary of E. Fermi.

BF2 Amaldi, Edoardo. "Enrico Fermi, premio Nobel 1938 per la fisica." Scienza e Tecnica, 2, no. 9-12 (1938): 284-88.

BF3 Feather, Norman. "Enrico Fermi, Hon. F. R. S. E., For Mem. R.S." (Obituary) Yearbook of the Royal Society of Edinburgh, 1954-55.

BF4 Fermi, Laura. Atomi in famiglia. Milano: Mondadori, 1954. Pp. 341. Translation: Atoms in the Family. My Life with Enrico Fermi. Chicago: University of Chicago Press, 1954. ix + 267 pp.

BF5 Latil, Pierre de. Enrico Fermi, Cristofor Columb al atomului. In românește de Petre Mihăileanu. București: Ed Ştiinţifică, 1965. 192 pp.

Presumably similar to his book Enrico Fermi: The Man and his Theories. New York: Eriksson, 1966.

BF6 Segré, Emilio. Enrico Fermi, Physicist. Chicago: University of Chicago Press, 1970. 276 pp. Translations: Enrico Fermi, fisico. Bologna: Zanichelli, 1971. 284 pp. Enrico Fermi - Fizik. Moskva: Mir, 1973. 323 pp.

Fletcher, H.

BF7 Gardner, Mark B. "Harvey Fletcher." (Obituary) Physics Today, 34, no. 10 (Oct. 1981): 116.

Fletcher, born in 1884, was known for research on the nature of speech and hearing and other aspects of acoustics.

Fock, V. HW

BF8 Veselov, M. G., P. L. Kapitza, and M. A. Leontovich. "Vladimir Aleksandrovich Fock (Obituary)." Soviet Physics Uspekhi, 18 (1976): 840-41.

Translated from Uspekhi Fizikcheskikh Nauk, 117 (1975): 375-76. He died in December 1974.

Fowler, R. @ HW

Frenkel, Y. I. @ HW

BF9 Frenkel, V. Ia. "Raboty Ia. I. Frenkeliîa po fizike atomnogo iadra." Materiali 4-ĭ Zimneĭ schkoly po teorii iadra i fizike bysokikh energii, part II. Leningrad (1969), pp. 296-312.

BF10 Frenkel, V. Ia. "The style of a scientific creative genius--Yakov Il'ich Frenkel (on the eightieth anniversary of his birth)." Soviet Physics Uspekhi, 17 (1975): 577-83.

Translated from Uspekhi Fizicheskikh Nauk, 113 (July 1974): 535-47.

BF11 Frenkel, V. Ia. (no title), article in Vospominaniia o Ia. I Frenkele. Leningrad: Nauka, 1976, pp. 241-270.

Friedman, A. A. @ HW

BF12 Gavrilov, A. F. "Vospominaniîa o Fridmane (K Perepiske A. A. Fridmana)." Trudy Instituta Istorii Estestvoznaniîa i Tekhniki AN SSSR, 22 (1958): 389-400.

Recollections of A. A. Friedman. Additions to the correspondence.

BF13 "Pis'ma A. A. Fridmana B. B. Golitsynu i V. A. Steklovu." Trudy Instituta Istorii Estestvoznaniîa i Tekhniki AN SSSR, 22 (1959): 324-388.

Letters fromA. A. Friedman to B. B. Golitsyn in the

years 1914-15, 1918-19. Publication and introductory essay by L. S.Polak; notes by A. F. Gavrilov.

BF14 Polubarinova-Kochina, P. Ia. "Aleksandr Aleksandrovich Fridman." Uspekhi Fizicheskikh Nauk, 80 (1963): 345-352.

Frisch, O HW

BF15 Bethe, H. A., and G. Winter. "Otto Robert Frisch." Physics Today, 33, no. 1 (Jan. 1980): 99-100.

BF16 Lemmerich, Jost. "Nachruf fuer Otto Robert Frisch." Physikalische Blaetter, 36 (1980): 43.

He died 21 September 1979; born in Vienna, 1904.

Froehlich, H.

BF17 Haken, H. "Herbert Fröhlich 70 Jahre alt." Physikalische Blaetter, 31 (1975): 664-665.

BF18 Haken, H., and M. Wagner, eds. Cooperative Phenomena. New York: Springer-Verlag, 1973. xv + 458 pp.

Festschrift for Herbert Froehlich, on his retirement from the Chair of Theoretical Physics at Liverpool; includes some biographical information and list of publications.

BG. Subjects: G

Gabor, D. $71 HW

BG1 Leith, E. N. "Dennis Gabor." (Obituary) Physics Today, 32, no. 6 (June 1979): 70-71.

Gamow, G.

BG2 Alpher, Ralph A., and Robert Herman. "Gamow dies; nuclear and astrophysicist was popular writer." Physics Today, 21, no. 10 (Oct. 1968): 102-3.

BG3 Reines, Frederick, ed. Cosmology, Fusion & Other Matters. George Gamow Memorial Volume. Boulder, Colorado: Colorado Associated University Press, 1972. Pp. xiv + 320.

Includes: R. A. Alpher and R. Herman, "Reflections on 'Big Bang' Cosmology," 1-14; Stanislaw M. Ulam, "Gamow - and Mathematics," 272-79; Max Delbrück, "Out of this World," 280-88; Leon Rosenfeld, "Nuclear Reminiscences," 289-99; Maurice M. Shapiro, "George Gamow - an

Appreciation," 300-303; Ralph A. Alpher and Robert Herman, "Memories of Gamow," 304-13.

Gerlach, W. HW

BG4 Berninger, E. H. "Walter Gerlach." (Obituary) Nachrichtenblatt der Deutschen Gesellschaft fur Geschichte der Medizin, Naturwissenschaft und Technik, 29 (1979), Heft 2. 110 pp.

BG5 Stierstadt, Klaus. "Walther Gerlach." Physikalische Blaetter, 36 (1980): 18-19.

He died 10 August 1979 at age 90.

Germer, L. H.

BG6 MacRae, A. U. (Obituary of Lester H. Germer) Physics Today, 25, no. 1 (January 1972): 93, 97.

(Germer died Oct. 3, 1971)

Gibbs, J. W. @ HW

BG7 Langer, R. E. "Josiah Willard Gibbs." American Mathematical Monthly, 46 (1939): 75-84.

BG8 Larmor, J. Obituary of J. W. Gibbs. Proceedings of the Royal Society of London (1904): 1-17.

BG9 Cesper, R. E. "Josiah Willard Gibbs." Journal of Chemical Education, 32 (1955): 267-268.

BG10 Rukeyser, Muriel. Willard Gibbs. Garden City, N.Y.: Doubleday, Doran, 1942. Reprinted, New York: E. P. Dutton, 1964. 465 pp.

BG11 Seeger, Raymond John. Men of Physics: J. Willard Gibbs, American Mathematical Physicist par excellence. Oxford & New York: Pergamon Press, 1974. xii + 290 pp.

Consists primarily of extracts from his writings.

BG12 Wilson, Edwin B. "The last unpublished notes of J. Willard Gibbs." Proceedings of the American Philosophical Society, 105 (1961): 545-558.

BG13 Wilson, E. B. "Josiah Willard Gibbs." American Scientist, 39 (1951): 287-289.

Comparison of Wheeler's biography with others.

BG14 Wilson, E. B. "Reminiscences of Gibbs by a student and colleague." Bulletin of the American Mathematical Society, 37 (1931): 401-16.

Goldhaber, M.

BG15 Feinberg, G., A. W. Sunyar, and J. Wesener, eds. A Festschrift for Maurice Goldhaber. (Transactions of the New York Academy of Sciences, series II, volume 40) New York: New York Academy of Sciences, 1980. xii + 294 pp.

Includes: "Preface" by the editors, vii-ix, and "Selected Publications of Maurice Goldhaber," 287-93.

Gombas, P. HW

BG16 Yutsis, A. P., and I. I. Glembotskii. "Pal Gombas, In memoriam." Soviet Physics Uspekhi, 15 (1973), 856-58. Translated from Uspekhi Fizicheskikh Nauk, 108 (1972): 605-7.

On the Hungarian physicist, who died in 1971.

Gorter, C. J. HW

BG17 Van den Handel, J. Obituary of C. J. Gorter. Physics Today, 33, no. 10 (Oct. 1980): 84.

Cornelius Jacobus Gorter died 30 March 1980 at age 73.

Goudsmit, S. A.

BG18 Goldhaber, Maurice. "Samuel A. Goudsmit." (Obituary) Physics Today, 32, no. 4 (April 1979): 71-72.

Goudsmit was born 11 July 1902 in The Hague, Netherlands; he died 4 December 1978 in Reno, Nevada.

Graaff, R. J. Van De @ HW

BG19 Burrill, E. S. "Van de Graaff, the man and his accelerators." Physics Today, 20, no. 2 (Feb. 1967): 49-52.

Green, M.

BG20 Raveche, H. J., ed. Perspectives in Statistical Physics. New York: North-Holland Pub. Co., 1981. xxvi + 366 pp.

Memorial to M. S. Green, with remarks on his work by E. G. D. Cohen and M. E. Fisher, curriculum vitae and list of publications.

Gregory, B.

BG21 _Bernard Gregory (1919-1977)_, Groupe des Publications, Document no. CERN/PU-ED 78-01, Organisation europeene pour la recherche nucleaire. Geneve: Aout, 1978. 19 pp.

Recollections of French physicist and CERN Director General Bernard Gregory by Charles Peyrou, John B. Adams, and Louis Leprince-Ringuet. Charles Peyrou's recollections include an account of cosmic ray work at Pic-du-Midi station and discovery of the K particle (meson).

Hahn, O. @ $44(Chemistry) HW CB

BH1 Hahn, Otto. _Dal radiotorio alla fissione dell'uranio. Autobiografia scientifica_. Torino: Boringhieri, 1968. 179 pp.

BH2 Amaldi, Edoardo. "Otto Hahn, premio Nobel per la chimica." _Ricerca scientifica e ricostruzione_, 6 (1945): 2-4.

BH3 Berninger, Ernst. _Otto Hahn_. Munich: Heinz Moos Verlag, 1969; Bonn-Bad Godesberg: Inter Nationes, 1970. 79 pp.

In English, with numerous illustrations.

BH4 Born, Hans-Joachim, et al. "Erinnerungen an Otto Hahn." _Nachrichten aus Chemie, Technik und Laboratorium_, 27, no. 7 (1979): 404-5.

Critique of F. Krafft's article with same title, followed by Krafft's "Replik" _ibid_., pp. 405-8.

Hartree, D. R.

BH5 Darwin, Charles. "Douglas Rayner Hartree, 1897-1958." _Biographical Memoirs of Fellows of the Royal Society_, 4 (1958): 103-16.

Heisenberg, W. $32 HW

BH6 Heisenberg, Werner. _Across the Frontiers_. Harper & Row: New York, 1974. 229 pp.

Includes "The Scientific Work of Albert Einstein," and "Planck's Discovery and the Philosophical Problems of Atomic Theory."

BH7 Hermann, Armin. "Werner Heisenberg. Eine Wuerdigung." Bild der Wissenschaft, 13, Heft 3 (1976): 52-57.

BH8 Hermann, Armin. "Abschied von Werner Heisenberg." Physikalische Blaetter, 32 (1976): 98-104.

BH9 Pfeiffer, Heinrich, ed. Denken und Umdenken zu Werk und Wirkung von Werner Heisenberg. Munich: Piper, 1977.

BH10 Thomsen, Dietrick. "An appreciation of Werner Heisenberg and some talk about how physics was in the good old days." Science News, 109, no. 10 (6 March 1976): 157.

Report on a press conference at which I. I. Rabi, S. Goudsmit and G. E. Uhlenbeck talked about Heisenberg. In reply to Goudsmit's statement that (in Nazi Germany) Heisenberg "saved physics, he did not save physicists," E. K. Gora later identified himself as "One Heisenberg did save," ibid, 109, no. 12 (20 March 1976): 179.

BH11 Wagner, Siegfried. "Lavoisier, Lichtenberg und Heisenberg: Revolutionaere der Wissenschaften ueber wissenschaftliche Revolutionen." Berichte zur Wissenschaftsgeschcichte, 4 (1981): 127-41.

Heisenberg's views on scientific revolutions.

BH12 Wigner, Eugene P. "Werner K. Heisenberg." (Obituary) Physics Today, 29, no. 4 (April 1976): 86-87.

Heitler, W.

BH13 Rasche, Günther. "Laudatio Professor Walter Heitler, anlässlich der Verleihung der goldenen Medaille der Humboldt-Gesellschaft (Aula der Universität Zurich 6 Mai 1979)." Archives Internationales d'Histoire des Sciences, 30 (1980): 162-66.

Hertz, G. $25 HW

BH14 Goudsmit, S. "Gustav Hertz." (Obituary) Physics Today, 29, no. 1 (Jan. 1976): 83-85.

BH15 "Gustav Hertz in der Entwicklung der modernen Physik." Abhandlungen der deutschen Akademie der Wissenschaften zu Berlin, Klasse für Mathematik, Physik und Technik, Jahrgang 1967, Nr. 1.

Festschrift for his 80th birthday (22 July 1967), with articles by W. Hartmann, R. Rompe and M. Steenbeck, etc.

Hevesy, G. @ $43(Chemistry) HW

BH16 Broda, Engelbert. "Georg von Hevesy 1885-1966." Chemie in unserer Zeit, 1 (1966): 73-75.

BH17 "Georg von Hevesy." (Obituary) Physics Today, 19, no. 8 (Aug. 1966): 95.

Hevesy won the Nobel Prize in Chemistry in 1943 for his work with radioactive tracers. He died 5 July 1966 at age 80.

Hueckel, E.

BH18 Suchy, K. "Erich Hueckel." Physics Today, 33, no. 5 (May 1980): 72-75.

Infeld, L. @ HW

BI1 Infeld, Erik. Leopold Infeld. Warszawa, 1978. 178 pp.

Ioffe, A. @ HW

BI2 Ioffe, A. F. Teoria și practica fizicii sovietice. București, 1950. 17 pp.

BI3 Ioffe, A. F. Vstrechi s Fizikami. Moi Vospominaniîa o Zarubezhnykh Fizikakh. Moskva: Fizmatgiz, 1960. 144 pp.

Recollections of A. Ioffe on Western and Soviet physicists.

BI4 Aleksandrov, A. P. "Hundredth Anniversary of the birth of A. F. Ioffe: Academician A. F. Ioffe and Soviet Science." Soviet Physics Uspekhi, 23(1981): 525-30. Frenkel, V. Ya., "Abram Fedorovich Ioffe (Biographical Sketch)." Ibid., 23 (1981): 531-50.

Both translated from Uspekhi Fizicheskikh Nauk, 132 (1980): 3-10, 11-45.

BI5 *Abram Fedorovich Ioffe*. (AN SSSR. Materialy k Biobibliografii Uchenykh SSSR. Seriiâ Fiziki, Vyp. 12) Introduction by A. I. Ansel'm and V. P. Zhuze, bibliography compiled by T. O. Vreden-Kobetskaya and E. I. Gusenkova. Moskva: Izd-vo AN SSSR, 1960. 135 pp.

Biobibliography of A. F. Ioffe.

BI6 Frenkel', V. Ia. "Kommentarii." *Izbrannie Trudy*, Vol. 2. By A. F. Ioffe. Leningrad: Nauka, 1975.

BI7 Frenkel, V. Ia. "Kommentarii." *O fizike i fizikakh*. By A. F. Ioffe. Leningrad: Nauka, 1977.

BI8 Frenkel, V. Ia. "A. F. Ioffe i sovetskaia fizicheskaia periodika." *Fizika Tverdogo Tela*, 22 (1980): 2881-5.

Ioffe's role in establishing Soviet physics journals.

BI9 Frenkel, V. Ia. "Akademik Ioffe. K stoletiiu so dnia rozhdeniia." *Vestnik AN SSSR*, 9 (1980): 105-109.

V. Ia. Frenkel, "A. F. Ioffe i N. Bor," ibidem, 116-121.

BI10 Frenkel, V. Ia and Ia. I. Frenkel. "Abram Fedorovich Ioffe." A. F. Ioffe, *Izbrannie Trudy*, Vol. I. Leningrad: Nauka (1974), pp. 7-26.

V. Ia. Frenkel, Commentaries, Ibidem.

BI11 Frenkel, V. Ia., and N. Ia. Moskovchenko. "Dokumenty i materialy (publikatsiia i kommentarii)." *Vklad Akad. A. F. Ioffe v stanovlenie i razvitie iadernoĭ fiziki v SSSR*. Leningrad: Nauka, 1980, pp. 8-38.

A collection of documents concerning Ioffe's role in establishing and developing nuclear physics in the USSR.

BI12 Frenkel V. Ia., and N. Ia. Moskovchenko. "Iz naslediîa akademika A. F. Ioffe." *Voprosy filosofii*, 12 (1980): 135-147.

BI13 Frenkel, V. Ia., and N. Ia. Moskovchenko. "Predislovie." *Nauchno-organizatsionnaia deiatel'nost' Akad. A. F. Ioffe*. Edited by L. S. Stil'bans and V. Ia. Frenkel. Leningrad: Nauka, 1980, pp. 3-10. With commentaries, ibidem.

BI14 Grigor'jan, A. T. "Abram Fedorovic Ioffe (1880-1960)." *NTM*, 18, no. 1 (1981): 44-49.

BI15 Ivanenko, D. D. "Abram Fedorovich Ioffe." _Osnovateli Sovetskoǐ Fiziki_. Moskva, 1970, pp. 103-134.

BI16 Kokin, L. "Etot Fantazer Ioffe..." _Puti v Neznaemoe Pisateli Rasskazyvaiut o Nauke_. Sbornik 7. Moskva, (1969), pp. 183-216.

On Ioffe the phantasist.

BI17 _Naucino-populiarna sesija v cest na Akademik Abram Feodorovic Ioffe (9.X.1952)_. Sofija, 1954. 66 pp.

BI18 Pisarzhevskiǐ, O. "Akademik Ioffe. Glavy iz Knigi." _Prometei_, 2 (1967): 34-58.

Academician Ioffe, Chapters from books.

BI19 Tuchkevich, V. M., and V. Ia. Frenkel. "FTI im. A. F. Ioffe v gody voiny." _Vestnik istorii estestvoznaniia i tekhniki_ (VIET), 2 (1975): 13-20.

BI20 "U Istokov Iadernoǐ Fiziki." _Vestnik AN SSSR_, 10 (1967): 81-87.

On the work of a "collective" of scientists in the '20s and '30s, with a report by I. V. Kurchatov "To Academician A. F. Ioffe. Illustration of the work of recent years," prepared by V. K. Voronovskiǐ and B. V. Levshin.

BI21 _Vospominaniia ob A. F. Ioffe_. Leningrad, Nauka, 1973. 250 pp.

Jensen, J. H. D. $63

BJ1 "Professor Dr. J. Hans D. Jensen." _Physikalische Blaetter_, 29 (1973): 233.

He was born 26 June 1907, died 11 February 1973. The announcement consists of a photograph and the statement "Eine Nachruf hat sich der Verstorbene ausdrücklich verbeten."

Joliot-Curie, F. @ $35(Chemistry) HW

BJ2 Biquard, Pierre. _Fréderic Joliot-Curie, The Man and his Theories_ (Profiles in Science). Greenwich: Fawcett Publications Inc., 1966. 192 pp.

A brief biography with selected writings and correspondence.

BJ3 Erenburg, I. G. *Frederik Zholio-Kiuri.* Moskva: Gospolitizdat, 1958. 31 pp.

BJ4 Gentner, Wolfgang. "Gespräche mit Frédéric Jolio-Curie im besetzten Paris 1940-1942." Heidelberg: Max-Planck-Institut für Kernphysik, 1980.

BJ5 Ghimesan, I. *Fr. Joliot Curie.*. Bucureşti: Ed. tineretului, 1961. 317 pp.

BJ6 Pontekorvo, Bruno. "Uchenyi, Borets, Chelovek." *Ogonek*, 35 (1958): 3.

Recollections of F. Joliot-Curie.

BJ7 Rouze', Michel. *Frederic Joliot-Curie.* Cu un cuvint inainte de Mihail Sadoveanu. Prefata de prof. Bernol. Bucuresti: Ed. de Stat pentru Literatura Stiintifica si Didactica, 1951. 81 pp.

BJ8 Šimane, Čestmír. *Frederic Joliot-Curie.* Prague: Horizont, 1980. 152 pp.

BJ9 Šimane, Čestmír. "Vzpomínka na Fréderica Joliot-Curie." *Pokroky matematiky, fyziky a astronomie*, 23 (1978): 301-7.

"Remembrance of F. J-C."

Joliot-Curie, I. *See* Curie, I.

Jordan, P. HW

BJ10 "Pascual, Jordan gestorben." *Physikalische Blaetter*, 36 (1980): 288.

Kapitsa, P. $78 HW

BK1 Kapitsa, P. L. _Collected Papers_. Edited by D. ter Haar. 3 vols. New York: Pergamon Pres, 1964-67. xvi + 503; viii + 505-993; x + 244 pp.

Includes item BL15 and BR17.

BK2 Kapitsa, P. L. _La scienza come impresa mondiale_. Roma: Editori Riuniti, 1979. 298 pp.

Includes: "Paul Langevin, fisico e personalita sociale," 179-180; "Aleksandr Aleksandrovich Fridman," 184-185; "Lev Davidovich Landau," 186-194; "Ricordi su Rutherford," 195-214.

BK3 Kapitsa, Peter Leonidovich. _Experiment, Theory, Practice: Articles and Addresses_. Hingham, Mass.: Reidel, 1980. xxvi + 424 pp.

Translation of second edition of _Eksperiment, teoriia, praktika_ (Moskva: Nauka, 1977). Includes a bibliography of his publications.

BK4 Kapitsa, P. L. _Eksperiment, teoriia, praktika_. (third edition completed by A. S. Borovik-Romanov and P. E. Rubinin) Moskva: Nauka, 1981. 495 pp.

Collection of non-technical papers by P. L. Kapitsa. Section II (129-165) deals with the beginnings and the organization of Moscow's Institute of Physical Problems; Section V (271-323) contains a series of recollections of E. Rutherford; Section VI (324-389) contains historical essays on Lomonosov and B. Franklin and recollections of Paul Langevin, A. Einstein, I. P. Pavlov, A. A. Friedman, L. D. Landau.

BK5 Alekseevskii, N. E. "Petr Leonidovich Kapitsa (On the occasion of his 70th birthday)." _Soviet Physics Uspekhi_, 7 (1965): 629-35. Translated from _Uspekhi Fizicheskikh Nauk_, 83 (1964): 761-68.

BK6 Biew, A. M. _Kapitsa. The story of the British-trained scientist who invented the Russian Hydrogen Bomb_. London: Frederick Muller Ltd., 1956. 288 pp.

BK7 Dobrovol'skiĭ, E. "Modus Vivendi. Dokum. Povest." _Puti V Neznaemoe_, Sb. 6, Moskva (1966): 83-137.

On the life and works of P. L. Kapitsa.

BK8 Dobrovol'skiĭ, E. N. Pocherk Kapitsy. Moskva: Sovetskaia Rossiia, 1968. 177 pp.

The handwriting of Kapitsa. Reviewed by V. Frenkel in Novyi Mir, 8 (1969): 284.

BK9 Frenkel, V. Ia. "Stat'i i rechi akad. P. L. Kapitsy." Uspekhi Fizicheskikh Nauk, 115, (1975): 335-339.

On articles and speeches by Academic Kapitsa.

BK10 Lombardo-Radice, Lucio. "Petr Leonidovich Kapitsa, scienziato umanista e rivoluzionario concreto."

Preface to P. L. Kapitsa, La scienza come impresa mondiale (item BK2), pp. 7-28.

BK11 Parry, A. The Russian Scientist. From Mendeleev and Pavlov to the brilliant scientists and technologists of today's USSR. New York: Macmillan, 1973. 196 pp.

Includes a chapter on "Peter Kapitsa: Faithful Sentinel of non-Marxist science," 115-38.

Karman, T. V. @ HW

BK12 Karman, Theodore von. Collected Works. Volume I, 1902-1913. London: Butterworths, 1956. ix + 531 pp.

This volume includes H. L. Dryden, "The Contributions of Theodore von Karman to Science and Technology," vii-ix, and reprints of papers on specific heats and vibrations of solids (with M. Born, 1912-13); also a comprehensive review on solids (with L. Foeppl).

BK13 Karman, Theodore von, with Lee Edson. The Wind and Beyond. Theodore von Karman. Pioneer in Aviation and Pathfinder in Space. Boston: Little, Brown, 1967. viii + 376 pp.

In addition to his major work in aerodynmics and aviation, von Karman worked with Max Born on the theory of crystal lattices, at Goettingen. The book includes "shrewd and amusing accounts of the German university community." (review in Science, 159 (1968): 517.

BK14 Hall, R. Cargill. "Theodore von Karman 1881-1963." Aerospace Historian, 28 (1981): 253-58.

BK15 (Obituary) Physics Today, 16, no. 7 (July 1963): 74.

He died May 6, 1963, less than a week before his 82nd birthday.

Klein, O. HW

BK16 Deser, Stanley. "Oskar Klein." (Obituary) Physics Today, 30, no. 6 (June 1977): 67-68.

He died 5 February 1977 at age 82.

Konstantinov, B. HW

BK17 Aleksandrov, A. P., L. A. Artsimovich, B. A. Gaev, Ya. B. Zeldovich, V. M. Tuchkevich, and V. Ya. Frenkel. "Boris Pavlovic Konstantinov." Soviet Physics Uspekhi, 13 (1970): 140-145.

Translated from Uspekhi Fizicheskikh Nauk, 100 (1970): 163-69.

BK18 Frenkel, V. Ia. "B. P. Konstantinov. Kratkiĭ ocherk nauchnoĭ, nauchno-organizatsionnoi i obshchestvennoĭ deiatel'nosti." Boris Pavlovich Konstantinov. Moskva: Nauka, 1976, pp. 6-17.

BK19 Frenkel, V. Ia., and V. Zaitsevoi. Boris Pavlovich Konstantinov 1910-1969. Moscow: Izdatelstvo "Nauka," 1976. 50 pp.

BK20 "Obituary of Soviet Physicist discloses key nuclear role." New York Times, 12 July 1969, p. 27.

Kowarski, L. HW

BK21 Kowarski, Lew. Reflexions sur la Science. Reflections on Science. Geneva: Institut Universitaire de Hautes Etudes Internationales, 1978. 272 pp.

A collection of popular articles, mostly in English, on nuclear energy, scientific institutions, and computers.

BK22 Goldschmidt, Bertrand L. "Lew Kowarski." (Obituary). Physics Today, 32, no. 12 (Dec. 1979): 68, 70.

BK23 Lew Kowarski (1907-1979), Groupe des Publications/PU-ED 80-07, Organization europeenne pour la recherche nucleaire. Geneve: Mai, 1980. 22 pp.

Recollections of Lew Kowarski by Jules Gueron, Charles Peyrou, Denis de Rougemont, Jean Mussard.

Kurchatov, I. @ HW

BK24 Artsimovich, L. A. "Bol'shoĭ Ucheniĭ."Kul'tura i Zhizn', 3 (1961): 20-21.

Recollections of I. V. Kurchatov.

BK25 Astashenkov, P. T. Podvig Akademika Kurchatov. Moskva: Znanie, 1979. 160 pp.

BK26 Golovin, I. N. "I. V. Kurchatov i Frederik Zholio-Kiuri." Istoriia i Metodologiia Estestvennykh Nauk, 6 (1968): 8-14.

On Kurchatov and Frederic Joliot-Curie.

BK27 Golovin, I. N. I. V. Kurchatov. Moskva: Atomizdat, (1967). 110 pp.

Reviewed by V. Frenkel. Zvezda, 1 (1968): 215-217.

BK28 "Igor Vasil'evich Kurchatov. Biographicheskii Ocherk." Uspekhi Fizicheskikh Nauk, 73, (1961): 593-604.

With a list of works of I. V. Kurchatov, 602-604.

BK29 Kikoin, I. K. "Igor' Vasil'evich Kurchatov." Atomnaîa Energiîa, 14, no. 1 (1963): 5-9.

BK30 Sokolov, Iu. L. "Iz Vospominaniĭ Fizika." Puti V Neznaemoe, Sb. 5, Moskva (1965): 460-489.

Recollections of I. V. Kurchatov.

Lampa, A. HW

BL1 Kleinert, Andreas. "Anton Lampa--ein Pionier der Hochfrequenzspektroscopie." Bibliothekarische Arbeit zwischen Theorie und Praxis. Festgabe für Wolfgang Thauer. Stuttgart, 1976, pp. 119-29.

Landau, L. D. @ $62 HW CB

BL2 Abrikosov, A. A. Akademik L. D. Landau. Kratkaîa Biografiîa I Obzor Nauch. Rabot. Moskva: Nauka (1965). 48 pp. (AN SSSR, Nauchno-Populiârnaiâ Seriîa).

Brief popular biography of L. D. Landau.

BL3 Andronikashvili, E. "Landau, kakim my ego Zapomnili." Puti v Neznaemoe. Pisateli Rasskazyvaîut o Nauke. Sbornik 8, Moskva (1970): 375-382.

"Landau as we knew him."

BL4 Bessarab, Maiia Gakhovlevna. Stranitsy zhizni Landau Moskva: Moskovskii Rabotsii, 1971. 134 pp. Translation: Stranky ze zivota L. D. Landaua. Prague: Mlada fronta, 1973. 118 pp.

BL5 Chernov, A., editor. _Nauchnoe Tvorchestvo L. D. Landau_. Moskva: Znanie, 1963. 31 pp.

Articles by V. L. Ginzburg and others on Landau's scientific work.

BL6 Grashchenkov, N. I., and N. N. Anichkov. "How the life of Academician L. D. Landau was saved." Livermore, Calif.: Lawrence Radiation Laboratory, report UCRL Trans. - 1018(L) (no date). Translated from _Priroda_, 52 (1963): 106-8.

BL7 Janouch, F. _Lev D. Landau: His Life and Work_. CERN 79-03, Geneva, 28 March 1979. 21 pp.

Colloquium given at CERN in June 1978 with a selected list of Landau's scientific papers and a list of books and articles about or by Landau.

BL8 Kompaneets, A. S. "L. D. Landau-Pedagog." _Priroda_, 4 (1969): 94-95.

BL9 "Lev Landau." _Ceskoslovensky casopis pro fyziku_, sekce A, 22 (1972): 284-89, 400-6.

BL10 Livanova, Anna. _Landau: A Great Physicist and Teacher_. New York: Pergamon Press, 1980. 217 pp.

Includes a substantial account of his research on the superfluidity of helium.

BL11 Tamm, I. E., Abrikosov, A. A., Khalatnikov, I. M. "L. D. Landau Laureat Nobelevskoi Premii 1962 Goda." _Vestnik AN SSSR_, 12 (1962): 63-67.

On Landau's 1962 Nobel Prize.

Lande, A. HW

BL12 Yourgrau, Wolfgang. "Alfred Lande." (Obituary) _Physics Today_, 29, no. 5 (May 1976): 82-83.

He died in Nov. 1975; born 13 December 1888.

Langevin, P. @ HW

BL13 Biquard, P. _Paul Langevin, scientifique, educateur, citoyen_. Paris: Editions Seghers, 1969. 196 pp.

BL14 "Le centenaire de Paul Langevin." Pensee, 165 (1972): 3-99.

Articles by H. Gratiot-Alphandery, L. Langevin, J. Beauvais, G. Snyders. See also J. Langevin, "Paul Langevin et le Journal de physique," ibid., 127-34.

BL15 Kapitsa, P. L. "The physicist and public figure Paul Langevin." Collected Papers of P. L. Kapitza (item BK1), pp. 208-214. (Text of lecture given in 1957)

BL16 Langevin, Andre. "O stycich Paula Langevina a jeho kolegu s ceskoslovenskymi fyziky." Dejiny ved a techniky, 10 (1977): 1-7.

Short biographical sketch and list of publications, followed by extracts from his writings.

BL17 Langevin, André. "Sur les relations de Paul Langevin et de ses collegues francais avec les physiciens tchecoslovaques." Acta Historiae rerum naturalium nec non technicarum, 10 (1977): 9-19.

Laue, M. V. @ $14 $32(Chemistry) HW

BL18 Hermann, Armin. "Max von Laue (1879-1960)." Vorbilder für Deutsche. Korrektur einer Heldengalerie. Edited by Peter Glotz and W. R. Langenbucher. Muenchen, 1974, pp. 122-38.

BL19 Herneck, Friedrich. "Max von Laue." Von Liebig zu Laue. Edited by Otto Finger and Friedrich Herneck. Berlin: VEB, Deutscher Verlag der Wissenschaften, 1963, pp. 345-354.

Lawrence, E. O. @ $39 HW

BL20 "Ernest Orlando Lawrence Papers." Bancroftiana (Univerity of California, Berkeley, Bancroft Library), no. 53 (Sept. 1972): 6-7.

Lazarev, P. P. @ HW

BL21 Kravets, T. P. "Tvorcheskiĭ Put' Akademika P. P. Lazareva." (1943). Ot N'iutona Do Vavilova Ocherki i Vospominaniia. Leningrad, 1967, pp. 328-337.

On the creative path of P. P. Lazarev.

BL22 Lazarev, P. P. "Iz Perepiski Akademika P. P. Lazareva. 1878-1942." Istoriîa i Metodologiîa Estestvennykh Nauk, 8 (1970): 223-249.

Extracts from the correspondence of P. P. Lazarev, with comments by O. A. Il'ina, A. F. Kononkov, and A. M. Tolmacheva.

BL23 Petr Petrovich Lazarev (1878-1942) (AN SSSR). Materialy k Biobibliographii Uchenykh SSSR. Seriia Fiziki, Vyr. 10) Introductory essay by M. P. Volarovich. Bibliography by N. M. Nesterova. Moskva: Izd-Vo AN SSSR (1958). 127 pp.

Biobibliography of P. P. Lazarev.

BL24 Radovskii, M. I. "Obraz Zamechatel'nogo Uchenogo. Tsennye Dokumenty O P. P. Lazareve." Priroda, 10 (1963): 94-96.

Documents on P. P. Lazarev.

Lebedev, A. A. HW

BL25 Aleksandr Alekseevich Lebedev. (AN SSSR. Materialy k Biobibliographii Uchenykh SSSR. Seriia Fiziki, Vyr. 8) Introductory essay by V. G. Vaphiadi, Bibliography by N. M. Nesterova. Moskva: Izd-Vo AN SSSR (1957). 27 pp.

Biobibliography of Alexandr Alekseevich Lebedev.

Lebedev, P. N. @ HW

BL26 Grigorian, A. T., ed. Trudy Instituta Istorii Estestvoznaniia i Tekhniki, T. 28, Istoriia Fiziko-Matematicheskikh Nauk. Moskva: Izd.-vo AN SSSR, 1959. 523 pp.

Recollections of P. N. Lebedev by P. P. Lazarev; T. P. Kravets; G. A. Shain; A. K. Timiriazev; V. K. Arkad'ev; N. A. Kaptsov; V. D. Zernov; S. N. Rzhevkin: A. I. Akulov; A. N. Amphiteatrova-Levitskaia.

BL27 Ivanenko, D. D., and P. I. Ziukov. "K Perepiske P. N. Lebedeva s B. B. Golitsynym." Istoriia I Metodologiia Estestvennykh Nauk 4 (1966): 307-309.

BL28 Petr Nikolaevich Lebedev. Bibliograficheskii ukazatel'. Sost. A. M. Lukomskaja. Pod red. K. I. Safranovskogo Moskva: Leningrad, 1950. 186 pp.

Lenz, W.

BL29 Jordan, P. "Wilhelm Lenz." Physikalische Blaetter, 13 (1957): 269-70.

Leontovich, M. A. HW

BL30 Kadomtsev, B. B. "Mikhail Alexandrovich Leontovich." (Obituary) Physics Today, 34, no. 12 (Dec. 1981): 62-63.

Lifshitz, I.

BL31 "Ilya Mikhailovich Lifshits (On his Sixtieth Birthday)." Soviet Journal of Low Temperature Physics, 3 (1977): 64-65. Translated from Fiz. Nizk. Temp. 3 (1977): 133-34.

Lindemann, F. See Cherwell, Lord.

Lorentz, H. @ $02 HW CB

BL32 Enskog, David. "Hendrik Antoon Lorentz, 18/7 1853--4/2 1928." Kosmos, 7 (1929): 5-15. In Swedish.

BL33 Frenk, A. M. "Voprosy Optiki i Teorii Izluchenii͡a v Rabotakh Lorentsa." Starye i Novye Problemy Fiziki (item BL35), pp. 302-326.

Problems of optics and radiation theory in Lorentz' works.

BL34 Haas-Lorentz, G. L. de, ed. H. A. Lorentz, Impressions of His Life and Work. North-Holland Publishing Company, Amsterdam, 1957. 172 pp.

Recollections of Lorentz by A. Einstein, W. J. de Haas, G. L. de Haas-Lorentz, A. D. Fokker, P. Ehrenfest, Balth van der Pol, H. B. G. Casimir.

BL35 Lorents, Gendryk Anton. Starye i Novye Problemy Fiziki. V. I. Frankfurt and L. S. Freĭman, eds. Moskva: Nauka, 1970. 370 pp.

Contains speeches by Lorentz and recollections of him, and the following supplement: V. I. Frankfurt & A. M. Frenk, "Raboty G. A. Lorentsa po Elektrodinamike i Termodinamike," pp. 287-301 and item BL 33.

Lukirskii, P. HW

BL36 Petr Ivanovich Lukirskiĭ. (AN SSSR. Materialy K Biobibliographii Uchenykh SSSR. Seriia Fiziki, Vyr. 11) Introductory essay by S. Iu. Luk'ianova and A. N. Murina. Bibliography by N. M. Nesterova, Moskva: Izd-Vo AN SSSR (1959). 41 pp.

Biobibliography of Peter Ivanovich Lukirskii (1894-1954).

Majorana, E. @ HW

BM1 Amaldi, Edoardo. "Nota biografica di Ettore Majorana." La vita e l'opera di Ettore Majorana. 1906-1938. Roma: Accademia Nazionale dei Lincei, 1966, pp. vii-xlix.

Personal and scientific biography of E. Majorana.

BM2 Amaldi, Edoardo. "Ettore Majorana, Man and Scientist." (Commemoration Speech) Strong and Weak Interactions-Present Problems. A. Zichichi, ed. New York: Academic Press, 1966, pp. 10-77.

BM3 Recami, Erasmo. "New Evidence about the Disappearance of the Physicist Ettore Majorana." Scientia, 110 (1975), 577-598; also in Tachyons, Monopoles and Related Topics, edited by E. Recami. Amsterdam: North-Holland, 1978, pp. 267-77.

Newly found letters and documents on Majorana's disappearance.

BM4 Sciascia Leonardo. La scomparsa di Ettore Majorana. Torino: Einauda, 1975. 77 pp.

Improbable novel of famous Italian writer on the disappearance of the physicist Ettore Majorana.

Mandel'shtam, L. I. @ HW

BM5 Leontovich, M. A., et al., eds. Akademik L. I. Mandel'shtam. K 100-letiiû so dnia rozhdeniia. Moskva: Nauka, 1979. 310 pp.

Contains a biography of L. I. Mandelshtam written by N. D. Papaleksi (5-52), correspondence of L. I. Mandelshtam (53-82), recollections of him by A. N. Krylov, G. S. Landsberg, A. A. Andronov, I. E. Tamm, G. S. Gorelik, S. M. Rytov, E. Ia. Shchegolev, V. L. Ginzburg and others. Also contains an early paper by L. I. Mandelshtam, and his essay on the optics of Newton and on A. N. Krylov.

BM6 Isakovich, M. A. "L. I. Mandel'shtam i rasprostranenie zvuka v mikroneodnorodnykh sredakh." Uspekhi Fizicheskikh Na[uk] 129 (1979): 531-40.

BM7 Rytov, S. M. "O Leonide Isaakoviche Mandel'shtame." Uspekhi Fizicheskikh Nauk, 129 (1979): 279-88.

BM8 Tamm, I. E. "Kharakternye Osobennosti Tvorchestva Leonida Isaakovicha Mandel'shtama," Uspekhi Fizicheskikh Nauk, 87 (1965): 3-7.

On the characteristics of the scientific work of L. I. Mandel'shtam.

Margenau, H.

BM9 Margenau, Henry. _Physics and Philosophy: Selected Essays_. Boston: Reidel, 1978. xxxviii + 404 pp.

Martinez-Risco, M.

BM10 Martinez-Risco, Manuel. _Oeuvres scientifiques_. Paris: Presses Universitaires de France, 1976. 250 pp.

Contains a "Noticia biografica" in Spanish on the life of Martinez-Risco (1888-1954), expert in optics, student of Blas Cabrera and P. Zeeman, who moved to Paris in 1936 to escape Franco's dictatorship. "Noticia biografica," 7-8; Rene Lucas, "Hommage a Martinez-Risco," 5-6.

Matthias, B. T.

BM11 Clogston, Albert M., Theodore H. Geballe, and John K. Hulm. "Bernd T. Matthias." (obituary) _Physics Today_, 34, no. 14 (Jan. 1981): 84.

BM12 Wohlleben, D., and C. Raub. "Bernd Matthias zum Gedenken." _Physikalische Blaetter_, 37 (1981): 125.

He was born in 1918, died 27 October 1980.

Mayer, M. G. $63 HW

BM13 Baranger, Elizabeth. "The present status of the nuclear shell model." _Physics Today_, 26, no. 6 (1973): 34-43.

Includes a personal note on Maria Mayer.

BM14 Sachs, Robert G. "Maria Goeppert Mayer, June 28, 1906-February 20, 1972." _Biographical Memoirs of the National Academy of Science_, 50 (1979): 311-28.

A revised version of this article appeared in _Physics Today_, 35, no. 2 (Feb. 1982): 46-51.

BM15 Wigner, Eugene P. "Maria Goeppert Mayer." (Obituary) _Physics Today_, 25, no. 5 (May 1972): 77, 79.

Meissner, W. HW

BM16 Brickwedde, F. G. "Walther Meissner." (Obituary) Physics Today, 28, no. 2 (February 1975): 84-85.

He died 16 November 1974, one month before his 92nd birthday.

Meitner, L. @ HW

BM17 Meitner, Lise. "Lise Meitner looks back." Advancement of Science, 20, no. 88 (1964): 39-46.

Comments on Boltzmann, Planck, Bohr and others.

BM18 Meitner, Lise. "Looking back." Bulletin of the Atomic Scientists, 20, no. 9 (Nov. 1964): 2-7.

BM19 Frisch, Otto R. "Lise Meitner dies; nuclear-physics pioneer." Physics Today, 21, no. 12 (Dec. 1968): 101.

Mendelssohn, K.

BM20 Kurti, N. "Kurt Mendelssohn." (Obituary) Physics Today, 34, no. 4 (April 1981): 87, 89.
He died 18 September 1980 at age 74.

Michelson, A. A. @ $07 HW CB

BM21 Jaffe, Bernard. Michelson and the Speed of Light. Garden City, NY: Doubleday, 1960. 197 pp.

Translation: Albert Michelson şi viteza luminii. (Translated by Z. Aurelia). Bucureşti: Ed. Ştiinţifică, 1967. 164 pp. (Savanţi de pretutindeni)

* Shankland, R. S., et. al. [Proceedings of the Michelson Colloquium in Potsdam, 28-29 April 1981] Cited below as item FA32.

Millikan, R. @ $23 HW

BM22 Goodstein, Judith, ed. The Robert Andrews Millikan Collection at the California Institute of Technology. Guide to a Microfilm Edition. Pasadena, CA: California Institute of Technology Archives, 1977. v + 98 pp.

BM23 Kargon, Robert. The Rise of Robert Millikan. Portrait of a Life in American Science. Ithaca, NY: Cornell University Press, 1982. 196 pp.

Møeller, C.

BM24 Goenner, Hubert. "Christian Møeller 1904-1980." Physikalische Blaetter, 36 (1980): 341.

Morley, E. W. @ HW

BM25 Williams, Howard R. Edward Williams Morley: His influence on science in America. Easton, Pa.: Chemical Education Pub. Co., 1957. xi + 282 pp.

Includes a brief section on the Michelson-Morley collaboration; Morley was hurt to be "summarily pushed aside" by Michelson, who ended the collaboration when Michelson moved to Clark University.

Morrison, P.

BM26 Eisenberg, Anne. "Philip Morrison -- A profile." Physics Today, 35, no. 6 (Aug. 1982): 36-41.

Includes recollections and comments on his involvement in the atomic bomb project, McCarthyism, astrophysics, and popular science writing.

Moseley, H. @ HW CB

BM27 Heilbron, J. L. H. G. J. Moseley: The Life and Letters of an English physicist, 1887-1915. Berkeley: University of California Press, 1974. xiii + 312 pp.

Comprehensive biography with many documents.

BM28 Trenn, T. J. "Moseley and more Moseleyana." Annals of Science, 33 (1976): 105-9.

Review of Heilbron's biography (item BM27) with two additional letters from Moseley to Bragg.

Mueller, E.

BM29 Tsong, T. T. "Erwin W. Mueller." (Obituary) Physics Today, 30, no. 8 (Aug. 1977): 70-71.

He was the first person to "see an atom" in 1955, using the field-ion microscope which he perfected.

Nagaoka, H. @ HW

BN1 Badash, Lawrence. "Nagaoka to Rutherford, 22 February 1911." Physics Today, 20, no. 4 (April 1967): 55-60.

Comments on the state of European physics in 1910.

BN2 Itakura, Kiyonobu, Tosaku Kimura and Eri Yagi. Nagaoka Hantara den. Tokyo, 1973. 719, 78 pp.

BN3 Yagi, Eri. "Research Group of the Committee for the Publication of Hantaro Nagaoka's Biography." _Japanese Studies in the History of Science_, 10 (1971): 23-24.

Nernst, W. @ $20(Chemistry) HW

BN4 Mendelssohn, Kurt. _Walther Nernst und seine Zeit. Aufstieg und Niedergang der deutschen Naturwissenschaften_. Weinheim: Physik-Verlag, 1976. 254 pp.

Translation from American edition.

Onsager, L. $68(Chemistry) HW

BO1 Montroll, Elliott W. "Lars Onsager" (Obituary) _Physics Today_, 30, no. 2 (Feb. 1977): 77.

Oppenheimer, J. R. @ HW CB

BO2 _Oppenheimer, J. Robert. A Register of His Papers in the Library of Congress_. Washington, D. C.: Library of Congress, Manuscript Division, 1974. 63 pp.

BO3 Arbab, John. "Oppenheimer and ethical responsibility." _Synthesis_, 5, no. 2 (1982): 22-43.

BO4 Chevalier, Haakon. _The Man who would be God_. New York: Putnam's Sons, 1959. 449 pp.

Novel based on the author's interactions with J. Robert Oppenheimer.

BO5 Goodchild, Peter. _J. Robert Oppenheimer: Shatterer of Worlds_. Boston: Houghton Mifflin, 1981. 302 pp.

BO6 Rabi, I. I., et al. _Oppenheimer_. New York: Scribner, 1969. x + 90 pp.

Includes speeches by R. Serber, V. F. Weisskopf, A. Pais, G. T. Seaborg and selected bibliography. Also published in _Physics Today_, 20, no. 10 (Oct. 1967).

BO7 Rouzé, Michel. _Robert Oppenheimer et la bombe atomique_. Paris: Seghers, 1962. 224 pp. Translations: _Robert Oppenheimer, The Man and His Theories_. Translated by Patrick Evans. New York: Eriksson, 1965 192 pp. _Robert Oppenheimer_. Translated by C. Popescu Ulmu. Bucuresti: Ed. stiintifica, 1967. 196 pp. (Savanti de pretutindeni)

BO8 Smith, A. K., and C. Weiner, eds. _Robert Oppenheimer: Letters and Recollections_. Cambridge, Mass.: Harvard University Press, 1981. 376 pp.

Papaleski, N. D. HW

BP1 Migulin, V. V. "N. D. Papaleski (k stoletiîu so dnia rozhdeniîa)." Uspekhi Fizicheskikh Nauk, 134 (1981): 519-26.

On Nikolai Dmitrievich Papaleski (1880-1947).

BP2 Salomonovich, A. E. "N. D. Papaleski i sovetskaîa radioastronomiîa." Uspekhi Fizicheskikh Nauk, 134 (1981): 541-50.

Pauli, W. @ $45 HW CB

BP3 Ehrenfest, Paul. "Address on Award of Lorentz Medal to Professor W. Pauli (October 31, 1931)." Collected Scientific Papers. Edited by Martin J. Klein. Amsterdam: North-Holland, 1959, pp. 617-22.

In German and Dutch.

BP4 Fierz, Markus, and V. F. Weisskopf, eds. Theoretical Physics in the Twentieth Century: A Memorial Volume to Wolfgang Pauli. New York: Interscience, 1960. x + 328 pp.

Includes R. Kronig, "The Turning Point" (attempts by the author, Pauli, and others to interpret atomic spectra, ca. 1925); W. Heisenberg, "Erinnerungen an die Zeit der Entwicklung der Quantenmechanik"; G. Wentzel, "Quantum theory of fields (until 1947)"; F. Villars, "Regularization and non-singular interactions in quantum field theory"; R. Jost, "Das Pauli-Prinzip und die Lorentz-Gruppe"; H. B. G. Casimir, "Pauli and the Theory of the Solid State"; M. Fierz, "Statistische Mechanik" (ergodic theory, ensembles, density fluctuations near critical point, etc.); V. Bargmann, "Relativity"; B. L. van der Waerden, "Exclusion principle and spin"; L. D. Landau, "Fundamental Problems"; C. S. Wu, "The Neutrino."

BP5 Hermann, Armin, K. v. Meyenn, and V. F. Weisskopf, eds. Wolfgang Pauli. Wissenschaftlicher Briefwechsel mit Bohr, Einstein,Heisenberg u. a.. Bd. I: 1919-1929, Springer Verlag, New York-Heidelberg-Berlin 1979. 577 pp.

BP6 Hermann, Armin. "Das Gewissen der Physik: Wolfgang Pauli." Bild der Wissenschaft, 5 (1980): 114-125.

* Herman, Armin. "Die Funktion von Briefen in der Entwicklung der Physik." Cited above as item B21.

BP7 Meyenn, Karl von. "Wolfgang Pauli 1900-1950." ETH-Bulletin der Eidgenossischen Technischen Hochschule Zurich, 154 (April 1980): 5-7.

BP8 Richter, Steffen. Wolfgang Pauli: Die Jahre 1918-1930. Skizzen zu einer wissenschaftlichen Biographie. Aarau/Frankfurt/Salzburg: Verlag Sauerlander, 1979. 112 pp.

Pauling, L. $54(Chemistry)

BP9 Pauling, Linus. "Fifty years of progress in structural chemistry and molecular biology." Daedalus, 99(1970): 988-1014.

Autobiographical notes, comments on early applications of quantum mechanics.

Peierls, R. HW

BP10 Aitchison, I. J. R., and J. E. Paton, ed. Rudolf Peierls and Theoretical Physics: Proceedings of the Symposium held in Oxford on July 11th & 12th, 1974, to mark the occasion of the retirement of Professor Sir Rudolph E. Peierls, F. R. S. C. B. E. Oxford: Pergamon Press, 1977. vii + 117 pp.

Perey, M.

BP11 Schofield, Maurice. "Women in the history of science." Contemporary Review, 210 (1967): 204-6.

The first woman member of the French Academy of Sciences was Mlle. Marguerite Perey, director of the nuclear research institute in Strasbourg and discoverer of Francium.

Perrin, J. @ $26 HW CB

BP12 Broglie, L. de, et al. "Celebration du centenaire de la naissance de Jean Perrin...1970." Revue d'Histoire des Sciences, 24, no. 2 (1971).

Persico, E.

BP13 Amaldi, Edoardo, and Franco Rasetti. "Ricordo di Enrico Persico (9 agosto 1900-17 giugno 1969)." Giornale di Fisica, Vol. 20, n. 4 (1979): 235-260.

Personal and scientific biography of E. Persico. Includes a full bibliography of Persico's works.

Placzek, G. HW

BP14 Amaldi, Edoardo. "George Placzek." La Ricerca Scientifica, 7 (1956): 2038-2042.

Commemoration of G. Placzek with a bibliography of his main works.

Planck, M. @ $18 HW CB

BP15 Hartmann, Hans. Max Planck als Mensch und Denker. Berlin: Siegismund, 1938. 189 pp. Second edition, Leipzig: Hirzel, 1948. 126 pp. Third edition, Basel: Ott, 1953. 255 pp.

BP16 Hermann, Armin. "Max Planck." Bild der Wissenschaft, (1969): 16-23.

BP17 Ioffe, A. F., and A. T. Grigorian, eds. Maks Plank, 1858-1958. Sbornik k stoletiiu so dnia rozhdeniia. Moskva: Izd-vo AN SSSR, 1958. 279 pp.

Includes bibliography of the basic works of M. Planck and the literature on him, compiled by M. G. Novlianskaia, pp. 245-77.

BP18 Kliaus, E. M., and U. I. Frankfurt. Maks Plank, (1858-1947) Moskva: Nauka, 1980. 392 pp.

The first part deals with the personal life of Max Planck and is written by E. M.Kliaus; the second deals with his scientific activity and is written by U. I. Frankfurt.

BP19 Kockel, B., W. Macke, and A. Papapetrou, eds. Max-Planck-Festschrift 1958. Berlin: VEB Deutscher Verlag der Wissenschaften, 1959. 413 pp.

Articles by H. Falkenhagen, H. Honl, K. Westpfahl, N. Bohr, V. Fock, L. de Broglie, L. Rosenfeld, P. A. M. Dirac, J. L. Destouches, L. Pauling, L. Janossy, P. Caldirola

Pomeranchuk, I. HW

BP20 Lapidus, L. I. "Isaac Iakovlevich Pomeranchuk. K Godovshchine so Dnîa Smerti." Priroda, 12 (1967): 51-54.

Pontecorvo, B.

BP21 Bogoliubov, N. N. "Laureat Leninskoĭ Premii B. M. Pontekorvo." Atomnaîa Energiîa, 14 (1963): 441-42.

On the occasion of Pontecorvo's Lenin prize, 1963.

Pontremoli, A.

BP22 Giordana, Gian Pietro. Vita di Aldo Pontremoli. Roma: Formiggini, 1933. 282 pp.

Biography of the Italian theoretical physicist who perished in the Nobile expedition to the North Pole.

Powell, C. F. @ $50 HW

BP23 Powell, Cecil Frank. Selected Papers. Edited by E. H. S. Burhop, W. O. Lock, and M. G. K. Menon. New York: American Elsevier, 1972. xiv + 455 pp.

Includes biographical information on Powell.

Rabi, I. $44 HW

BR1 Bernstein, Jeremy. Experiencing science: Profiles in Discovery. New York: Basic Books, 1978. 275 pp.

Includes a long profile of I. I. Rabi.

BR2 Motz, Lloyd, ed. A Festschrift for I. I. Rabi. New York: New York Academy of Sciences, 1977. x + 244 pp. (Transactions, Series II, vol. 38).

Includes historical articles by S. Millman, V. W. Hughes, W. E. Lamb, Jr., N. F. Ramsey, and A. Pais.

Resibois, P.

BR3 Prigogine, I., and G. Dewel. "Pierre Resibois: 1936-1979," Journal of Statistical Physics, 24 (1981): 7-19.

Discusses especially his work on non-equilibrium statistical mechanics.

Roentgen, W. C. @ $01 HW CB

BR4 Tomas, V. K. "Tri Pis'ma Russkikh Fizikov V. K. Rentgenu."Uspekhi Fizicheskikh Nauk, 90, (1966): 541-44. Translation: "Three letters of Russian physicists to W. K. Roentgen." Soviet Physics Uspekhi, 9 (1967): 913-15.

Letters of P. N. Lebedev, I.I. Borgman and O.D. Khvol'son to W. C. Roentgen, Jan.-Feb. 1896.

Rojansky, V.

BR5 Goble, Alfred T. "Vladimir Rojansky." (Obituary) Physics Today, 34, no. 8 (Aug. 1981): 76.

Rojansky, who died 4 March 1981 at age 80, was known primarily as a teacher and for his introductory quantum mechanics textbook

Rosenfeld, L. HW

BR6 Rosenfeld, Leon. Selected Papers of Leon Rosenfeld. Edited by R. S. Cohen and J. J. Stachel. (Boston Studies in the Philosophy of Science, Volume XXI) Boston: Reidel, 1979. xxxiv + 929 pp.

Includes S. Rozental, "Introduction"; L. Rosenfeld, "My initiation"; R. S. Cohen & J. J. Stachel, "Editorial Preface"; bibliography of Leon Rosenfeld (1904-1974).

BR7 Bohr, A. "Leon Rosenfeld." (Obituary). Physics Today, 27, no. 6 (June 1974): 71-72.

He died on March 23, 1974.

BR8 Prigogine, I. "Leon Rosenfeld et les fondements de la physique moderne." Bulletin de la Classe des Sciences, Academie Royale de Belgique, 5e Serie, 60 (1974): 841-54.

Rozhdestvenskii, D. S. HW

BR9 [Rozhdestvenskii, D. S.] "Iz Perepiski D. S. Rozhdestvenskogo c inostrannymi Fizikami." Istoriia I Metodologiia Estestvennykh Nauk, 3 (1965): 305-327.

Publication of letters from Academy Science Archives prepared by O. A. Il'ina, A. F. Kononkov, A. N. Osinovskii, A. M. Tolmacheva. Correspondence of D. S. Rozhdestvenskii with P. Drude, R. Wood, P. Ehrenfest, H. A. Lorentz, A. Smekal, Niels Bohr, F. Paschen, R. Ladenburg.

BR10 Gulo, D. D., and A. N. Osinovskii. Dmitrii Sergeevich Rozhdestvenskii. Moskva: Nauka, 1980. 283 pp.

Includes a list of works by D. S. Rozhdestvenskii (276-279) and a list of books and articles about him (280-283). A. N. Osinovskii is the author of the Introduction and the biographical chapters 1-12; more technical chapters 13-16 on R.'s scientific work are written by D. D. Gulo.

Rutherford, E. @ $08(Chemistry) HW CB

BR11 Andrade, E. N. Da Costa. Rutherford and the Nature of the Atom. Garden City, NY: Doubleday, 1964. xix + 218 p. Translation: Rutherford, come si scopri la natura dell'atomo. Bologna: Zanichelli, 1967. 183 pp.

BR12 Bunge, Mario & W. R. Shea, eds. Rutherford and Physics at the Turn of the Century. New York: Science History Publications, 1979. 184 pp.

Articles by E. N. Hiebert, L. Badash J. L. Heilbron, N. Feather, T. J. Trenn, S. L. Jaki, N. Cameron, S. G. Brush and G. Righini.

BR13 Danin, D. "Admiral Byl Velikodushen." Puti V Neznaemoe. Sb. 5, Moskva (1965): 409-452.

On the friendship between E. Rutherford and F. Soddy, 1898-1908.

BR14 Feather, Norman. "Ernest Rutherford." Nobel Prize Winners. Edited by L. J. Ludovici. London: Arco, 1957, pp. 15-33.

BR15 Hoyer, Ulrich. "Ernest Rutherford." Die Grossen der Weltgeschichte. Edited by Kurt Fassmann et al. Bd. 9. Munchen: Zurich, 1970, pp. 815-25.

BR16 Kapica, P. L. "Vzpominky na lorda Rutherforda." Československy časopis pro fyziku, sekce A, 20 (1970): 59-65, 181-85.

BR17 Kapitza, P. L. Collected Papers (item BK1). Volume III.

Includes reprints of four articles by Kapitza on Rutherford, from Nature (1937), Uspekhi Fizikcheskikh Nauk (1938), Bolshaya Sov. Enc. (1941) and Proceedings of the Royal Society of London (1966).

BR18 Kapitsa, P. L. "Moi Vospominaniia O Rezerforde," Novyi Mir, 8 (1966): 205-215.

Recollections of Rutherford.

Sackur, O.

BS1 Obituary of Otto Sackur. Nature, 94 (1914): 486.

Saha, M. N. @ HW

* Anderson, Robert S. Building Scientific Instiiutions in India: Saha and Bhaba. Cited below as item C3.

Sakharov, A.

BS2 Sakharov, Andrei. "The social responsibility of scientists." Physics Today, 34, no. 2 (June 1981): 25-30.

Includes "An autobiographical Note."

Satterly, J.

BS3 Satterly, John. "Reminiscences in physics from 1894 onward." American Journal of Physics, 25 (1957): 288-300.

On the spinthariscope, Lord Kelvin, Rutherford, J. J. Thomson, and A. Eddington.

Saulnier, G.

BS4 Nickles, H. "Gaston Saulnier, 1898-1974." Revue d'acoustique, 7, no. 29 (1974): 57.

Scherrer, P. HW

BS5 Frauenfelder, H., O. Huber, and P. Staehelin, eds. Beiträge zur Entwicklung der Physik. Festgabe zum 70. Geburtstag von Professor Paul Scherrer. Basel & Stuttgart: Birkhäuser Verlag, 1960. 253 pp.

Contains: P. Debye, "Paul Scherrer und die Streuung von Roentgenstrahlen," 9-13; essays and recollections by C. G. Suits, Gerhard Herzog, Rudolf Stössel, Hans H. Staub. Biographical Note on Paul Scherrer, p. 253.

Schiff, L. I.

BS6 Bloch, F., and J. D. Walecka. "Leonard I. Schiff." (Obituary) Physics Today, 24, no. 7 (July 1971): 54.

Schottky, W.

BS7 Spenke, E. "Schottky W. 1886-1976" (Nachruf), Physikalische Blaetter, 32 (1976): 170-2.

Schroedinger, E. @ $33 HW

BS8 Dirac, P. A. M. "Professor Erwin Schroedinger, For. Mem. R. S." (Obituary) Nature, 189 (1961): 355-56.

"He has told me how he came to make his great discovery..."

Simon, F. E. @

BS9 Kurti, N. "Franz Eugen Simon 1893-1956." Biographical Memoirs of Fellows of the Royal Society, 4 (1958): 225-56.

Skobel'tsyn, D. HW

BS10 Basov, N. G., S. N. Vernov, and A. I. Isakov. "Dmitrii Vladimirovich Skobel'tsyn (on his eightieth birthday)." Soviet Physics Uspekhi, 15 (1973), 859-60. Translated from Uspekhi Fizicheskikh Nauk, 108 (1972), 771-72.

He is known for work on interaction of gamma rays with matter, discovery of cosmic ray charged particles and showers.

BS11 Veksler, V. I. "Dmitriĭ Vladimirovich Skobel'tsyn. K 70-Letii͡u so dni͡a Rozhdenii͡a," Uspekhi Fizicheskh Nauk, 78 (1962): 539-544.

On the 70th birthday of D. V. Skobel'tsyn.

Slater, J. C. HW

BS12 Andre, Jean-Marie. "John Clark Slater, sa vie et son oeuvre Scientifique." Revue des Questions Scientifiques, 148 (1977): 445-56.

BS13 Kimball, G. E. "Reminiscences of 1933-35." International Journal of Quantum Chemistry, Symposium No. 1 (1967): 845-47.

Personal anecdotes about Slater's group at M.I.T.

BS14 Morse, Philip M. "John Clarke Slater, December 22, 1900 - July 25, 1976." Biographical Memoirs, National Academy of Sciences, 53 (1982): 297-321.

In 1929 Slater introduced the determinant method for constructing antisymmetric wave functions of electrons in atoms. He pioneered the applications of quantum mechanics to molecules and solids.

BS15 Slater, John C. Solid-State and Molecular Theory, A Scientific Autobiography. New York: Wiley, 1975. 357 pp.

An autobiographical account of "wave mechanics in the classical decade, 1923-1932," and subsequent decades when the author pioneered the application of quantum theory to solid state physics and quantum chemistry.

BS16 Van Vleck, John H. "John C. Slater." (Obituary) Physics Today, 29, no. 10 (October 1976): 68-69.

Smoluchowski, M. @ HW

BS17 Teske, A. "An outline account of the work of Marian Smoluchowski 1872-1917." Monografie Dziejow Nauki i Techniki, 51 (1970): 14-20.

BS18 Ulam, S. "Marian Smoluchowski and the theory of probabilities in physics." American Journal of Physics, 25 (1957): 475-81.

Sommerfeld, A. @ HW CB

BS19 Sommerfeld, Arnold. Gesammelte Schriften. Braunschweig: Vieweg, 1968. 4 Volumes.

Band IV inludes "Die Entwicklung der Physik in Deutschland seit Heinrich Hertz." (1918), "Zwanzig Jahre spectroscopischer Theorie in Muenchen" (1942), "Philosophie und Physik seit 1900." (1948), "Fifty Years of Quantum Theory" with F. Bopp (1951), and "Autobiographische Skizze." (1959).

BS20 Bopp, Fritz, and H. Kleinpoppen, eds. Physics of the One-and-Two-Electron Atoms. New York: American Elsevier, 1969. 873 pp.

Proceedings of Sommerfeld Centennial Memorial Meeting, Munich, 1968. Articles by P. P. Ewald, A. Hermann, B. L. van der Waerden, W. Heisenberg, H. A. Bethe and others.

BS21 Hermann, Armin. "Arnold Sommerfeld." Die Grossen der Weltgeschichte. Bd. 9 Zurich : Kindler, 1970, pp. 703-715. Ibid in Die Ludwig-Maximilians-Universität Muenchen in ihren Fakultaeten. Bd. 1. Berlin : Duncker und Humblot, 1972, pp. 435-451.

Strassman, F.

BS22 Friedlander, G., and G. Herrmann. "Fritz Strasmannn." (Obituary) Physics Today, 34, no. 4 (April 1981): 84-86.

He died 22 April 1980 at age 78.

BS23 Krafft, Fritz. Im Schatten der Sensation: Leben und Wirken von Fritz Strassmann nach Dokumenten und Aufzeichungen Dargestellt. Weinhem, West Germany & Deerfield Beach, Florida: Verlag Chemie, 1981. xvii + 541 pp.

Tamm, I. @ $58 HW

BT1 Igor Evgen'evich Tamm. (AN SSSR. Materialy k Biobibliographii Uchenykh SSSR. Seriia Fiziki, Vyp. 9). Introductory essay by V. L. Ginzburg and E. L. Feinberg. Bibliography by N. M. Nesterova. Moskva: Izd-Vo AN SSSR, 1959. 36 pp.

Biobibliography of I. E. Tamm up to 1953.

BT2 Bolotovsky, B. "Igor Tamm." (Obituary) Physics Today, 24, no. 9 (Sept. 1971): 71, 73.

He died 12 April 1971 at age 75.

BT3 Ginzburg, V. L., M. A. Markov, A. D. Sakharov, and E. L. Feǐnberg. "Pamiati Igoria Evgen'evicha Tamma," Uspekhi Fizicheskikh Nauk, 105 (1971): 163-164.

BT4 Ginzburg, V. L., I. M. Dremin, M. A. Markov, E. L. Feǐnberg, and V. Ia. Frenkel, eds. Vospominaniia o I. E. Tamme. Moskva: Nauka, 1981. 296 pp.

Contains recollections of the Nobel Prize Winner I. E. Tamm (1895-1971), with essays by S. A. Al'tshuler, E. L. Andronikashvili, E. S. Billig, B. M. Bolotovskiǐ, S. V. Vonsovskiǐ, V. L. Ginzburg, D. S. Danin, E. K. Zavoǐskiǐ, B. Kuznetsov, R. E. Peierls, N. V. Timofeef-Resovskiǐ, I. M. Fran[k], V. Ia. Frenkel', and others.

Contains a brief biographical sketch by V. L. Ginzburg and E. L. Feǐnberg, pp. 5-15; an essay by a relative of Tamm, [..] I. Vernskiǐ, based on family archives, and personal recollections, etc. (pp. 61-103)

V. Ia. Frenkel's contribution is an essay entitled "Vstrec[hi]" (Encounters), pp. 256-289, with his personal recollections of Tamm and of his meeting with other physicists (Dirac etc.). Frenkel also contributed an appendix (pp. 283-287) that has the reproduction of three letters from Tamm to P. Ehrenfest, and also excerpts of letters from Tamm to Sarra Isaakovna Frenkel', the widow of Yakov I. Frenkel (and mother of Viktor Iakovlevich Frenkel).

Taub, A.

BT5 Tipler, Frank J., ed. Essays in General Relativity. A Festschrift. New York: Academic Press, 1980. xvii + 236 pp.

Includes a poem "To Abe Taub, The Universe Man" by J. A. Wheeler; curriculum vitae and list of publications; photograph.

Taylor, G. HW

BT6 Pippard, Brian A. Obituary of Sir Geoffrey Taylor. *Physics Today*, 28, no. 9 (Sept. 1975): 67.

Teller, E. HW

BT7 Mark, Hans, and Sidney Fernbach, eds. *Properties of Matter Under Unusual Conditions*. In honor of Edward Teller's 60th birthday. New York: Interscience, 1969. ix + 389 pp.

Includes an article on Teller by E. P. Wigner, and "The concept of 'understanding' in theoretical physics" by W. Heisenberg.

BT8 Vilenskiĭ, M. E. *Vash Vrag Teller*. Moskva: Politizdat, 1964. 40 pp.

"Your enemy Teller." Pamphlet in the series "Vot oni 'beshenye'" ("here are the 'furious' ones").

Thomson, J. J. @ $06 HW CB

BT9 Rayleigh, Lord. "Sir Joseph John Thomson (1856-1940)." *Obituary Notices of Fellows of the Royal Society of London*, 3 (1941): 587-609.

BT10 Thomson, George Paget. *J. J. Thomson and the Cavendish Laboratory in his Day*. London: Nelson, 1964. 186 pp. Reprinted as *J. J. Thomson, Discoverer of the Electron*. Garden City, N. Y.: Doubleday/Anchor, 1966. 215 pp.

Tomonaga, S. $65 HW

BT11 Hayakawa, Satio. "Sin-intiro Tomonaga." (Obituary) *Physics Today*, 32, no. 12 (Dec. 1979): 66, 68.

He was born in 1906, died 8 July 1979.

Touschek, B.

BT12 Amaldi, Edoardo. *The Bruno Touschek Legacy (Vienna 1921--Innsbruck 1978)*. Geneva: Organisation Européenne pour la Recherche Nucléaire (CERN report 81-19), 1981. 83 pp.

"A biographical portrait of Bruno Touschek, an Austrian physicist who, before and during the Second World War, went through the dramatic adventures of a non-pure arian young person. Later he worked in Germay, Great Britain and Italy. Touschek was the first to propose chiral symmetry. He was the initiator and main driving force in the early developments of e^+e^- colliding machines..." (Author's abstract)

Truesdell, C.

BT13 Coleman, B., and J. Serrin, eds. Archive for Rational Mechanics and Analysis, 70 (1979).

This volume is dedicated to Clifford Ambrose Truesdell on his 60th birthday. Includes a list of his published works.

Tuve, M. A.

BT14 Gargan, Edward A. "Merle Tuve dies; aided in radar development." New York Times, 22 May 1982, p. 14.

Obituary of Merle A. Tuve (1901-1982).

Urey, H. C. $34(Chemistry)

BU1 Cohen, Karl. "Harold C. Urey 1893-1981." Bulletin of the Atomic Scientists, 37, no. 5 (May 1981): 8, 54-56.

BU2 Craig, H., S. L. Miller, and G. J. Wasserburg, eds. Isotopic and cosmic Chemistry. Amsterdam: North-Holland, 1964. xxv + 553 pp.

Dedicated to Harold Clayton Urey on his seventieth birthday, April 29, 1963. Includes recollections by R. M. Hutchins, J. H. Hildebrand, and the editors; article on discovery of deuterium by G. M. Murphy.

BU3 Garfield, E. " A tribute to Harold Urey." Current Contents--Physical, Chemical & Earth Sciences, 3 (December 1979): 5-9.

BU4 Stuckey, William. "Urey disclosed as key selector of Nobel winners." New York Times, 7 December 1975, p. 64.

BU5 Thode, H. G., and H. Alfven. "Harold Urey." (Obituary) Physics Today, 34, no. 4 (April 1981): 82, 84.

Van den Broek, A. J.

BV1 Elyashevich, M. A., and Y. I. Lisnevsky. "New materials on the life and scientific activities of A. J. van der Broek." Janus, 68 (1981): 241-79.

Van Vleck, J. H. $77 HW

BV2 Van Vleck, J. H. "My Swiss visits of 1906, 1926, and 1930." Helvetica Physica Acta, 41 (1968): 1234-7.

BV3 Anderson, Philip W. "John Hasbrouck Van Vleck." (Obituary) Physics Today, 343, no. 1 (Jan. 1981): 85,87.

Vavilov, S. @ HW

BV4 Frank, I. M., A. V. Shubnikov, B. A. Vvedenskii, A. L. Mints, S. N. Rzhevkin, N. N. Malov, G. P. Faerman, A. A. Lebedev, and N. A. Smirnova. "Sketches for a portrait of S. I. Vavilov." Soviet Physics Uspekhi, 16 (1974): 702-11; 17 (1975): 950-62. Translated from Uspekhi Fizicheskikh Nauk, 111 (1973): 173-90; 114 (1974): 53-54.

BV5 Frank, I. M., ed. Sergei Ivanovich Vavilov. Ocherki i vospominaniia. Moskva: Nauka, 1979. 296 pp.

Essays and recollections of S. I. Vavilov (1891-1951) by I. M. Frank, E. V. Shpol'skii, V. L. Levshin, A. N. Terenin, A. V. Shubnikov, G. S. Landsberg, V. Ronchi, V. I. Veksler, A. L. Mints, P. A. Cherenkov, N. A. Smirnov. Contains also the beginning of an autobiography by S. I. Vavilov (79-101) and a report on a trip he made to Italy (Verona and Arezzo) in 1913 (102-111).

BV6 Keler, Vl. R. Sergei Vavilov. Moskva, 1975. 317 pp.

Veksler, V. @ HW

BV7 Lobanov, Iu. N. "Akademik Vladimir Iosifovich Veksler," (Obituary) Fizika V Shkole, 3 (1966): 19-21.

BV8 Sarantseva, V. "Kogda Rozhdaîutsîa Idei?," Kul'tura I Zhizn', 5 (1966): 42-44.

On the works of V. I. Veksler (1907-1966).

BV9 Semenîushkin, I. N.,I. V. Chuvilo. "V. I. Veksler - Laureat Premii 'Atom Dlîa Mira'," Priroda, 10 (1963): 108-9.

Waldmann, L.

BW1 Hess, Siegfried, and Walter Köhler. "Ludwig Waldmann zum Gedenken." Physikalische Blaetter, 36 (1980): 155-6.

Webster, D. L.

BW2 Kirkpatrick, Paul. "David Locke Webster II, November 6, 1888 - December 17, 1976." *Biographical Memoirs, National Academy of Sciences*, 53 (1982): 367-400.

Weisskopf, V. F.

BW3 Weisskopf, Victor F. "My life as a physicist." *Physics in the Twentieth Century* (item A141), pp. 1-21.

Wigner, E. $63 HW

BW4 Wigner, Eugene. "Seeking to recognize order in nature." *Bulletin of the International Atomic Energy Agency*, 10 (1968): 38-44.

Wu, C. S.

BW5 Lubkin, Gloria. "Chien-Shiung Wu, the first lady of physics research." *Smithsonian*, 1, no. 10 (Jan. 1971): 52-57.

Yalow, R. $77(Physiology/Medicine)

BY1 Yalow, Rosalyn. "A physicist in biomedical investigation." *Physics Today*, 32, no. 10 (Oct. 1979), 25-29.

Review of her work leading to the 1977 Nobel Prize.

Yukawa, H. $49 HW

BY2 Brown, Laurie M. "Hideki Yukawa." (Obituary) *Physics Today*, 35, no. 2 (Feb. 1982): 88-89.

Yukawa died 8 September 1981, born 1907.

Zemansky, M. W.

BZ1 Hofstadter, Robert, H. Lustig, and H. Semat. "Mark W Zemansky." (Obituary) *Physics Today*, 35, no. 3 (March 1982): 73-74.

Zemansky was born in 1900, died 29 December 1981.

C. SOCIAL AND INSTITUTIONAL HISTORY

C1 Akademiia nauk Soiuza SSR, 220 let, 1725-1945, Ocherki po istorii Akademii nauk. Fizikomatematicheskie nauki. Moskva: Leningrad, 1945. 76 pp.

C2 Amaldi, Edoardo. "The Years of Reconstruction." Scientia, 114 (1979): 51-68, 439-51. Also in Perspectives of Fundamental Physics (item BA3), 379-461. Italian version in Giornale di Fisica, 20 (1979): 186-225.

A personal account of Italian physics in the years 1938-54.

C3 Anderson, Robert S. Building scientific institutions in India, Saha and Bhabha. Montreal: McGill University Centre for Developing-Area Studies, 1975. 124 pp.

C4 Auger, Pierre. "The Founding of CERN - Letter to the Editor." Bulletin of the Atomic Scientists, 12 (1956): 232.

C5 Baretl, H., W. Kalweit, and G. Kröber, Eds. Verbündete in der Forschung. (Traditionen der deutsch-sowjetischen Wissenschafts-Beziehungen und die wissenschaftliche Zusammenarbeit zwischen der Akademie der Wissenschaften der UdSSR und der Akademie der Wissenschaften der DDR) Berlin: Internationalen Reihe des Zentralinstituts für Geschichte der Akademie der Wissenschaften der DDR, 1976. 317 pp.

Contains Schlicker W., "Max Planck und die deutsch-sowtische Akademiebeziehungen während der Weimarer Republik."

C6 Belloni, Lanfranco. "Il Congresso di Como del 1927 vissuto e descritto da un ospite dell'Unione Sovietica." Museoscienza, 4 (ott.-dic. 1981): 14-21.

Italian translation from Russian of a report by Jacob (Yakov) Frenkel about the 1927 physics conference at Como.

C7 Beyerchen, Alan D. Scientists under Hitler: Politics and the Physics Comunity in the Third Reich. New Haven: Yale University Press, 1977. 287 pp. Translation: Wissenschaftler unter Hitler. Physiker im Dritten Reich. Koln: Kiepenheuer u. Witsch, 1980. 379 pp.

C8 Bird Dogs, The. "The evolution of the Office of Naval Research." Physics Today, 14, no. 8 (Aug. 1961): 30-35.

C9 Biriukov, V. A., M. M. Lebedenko, and A. M. Ryzhov. *Dubna, 1956-1966*. Dubna: Izd. OIIaI 1966. 274 pp.

C10 Blokhintsev, D. I. "Desiat' Let Nauchnoi Raboty Ob'edinennogo Instituta Iadernykh Issledovanii." *Atomnaia Energiia*, 20, no. 4 (1966): 293-310.

Ten years of scientificwork at the All Union Institute of Nuclear Research.

C11 Bloland, Harland G. and Sue M. Bloland. *American Learned Societies in Transition*. New York: McGraw-Hill, 1974. xvii + 130 pp.

Chapter 6, "Liberal reform in the American Physical Society." (on gerontocracy after 1945).

C12 Bogoliubov, N. N. "Ob'edinennomu Institutu Iadernykh Issledovanii v Dubne - 10 Let." *Prirody*, 6 (1966): 19.

On the tenth anniversary of the Dubna laboratory.

C13 Buchheim, Gisela. "Initiativen zur Gruendung der Physikalisch-Technischen Reichsanstalt (1887)." *NTM*, II (1974): 33-43.

C14 *CERN Courier*, 25th Anniversary Issue, vol. 19, no. 6, (September 1979).

Includes: " A brief history of CERN," 228-32; "The 25th Anniversary Ceremony," 233-38; "CERN physics, past and future," 239-41.

C15 "Deux initiatives du CEC: Documents sur les Origines du CERN et de la Fondation Europeenne de la Culture." *Centre Europeen de la Culture*, no. 4 (Hiver 1975). 68 pp.

Contains "note liminaire" and "Prehistoire due CERN (Itineraire d'une idee I-IV)."

C16 Cochrane, Rexmond C. *Measures for Progress: A History of the National Bureau of Standards*. Washington, D.C.: U. S. Department of Commerce, National Bureau of Standards, 1966. xxv + 703 pp.

Includes chapters on Samuel Wesley Stratton, Edward B. Rosa, Herbert Hoover, George Kimball Burgess, Lyman James Briggs, Edward Uhler Condon.

C17 Cochrane, Rexmond C. The National Academy of Sciences. The First Hundred Years 1863-1963. Washington, D. C.: National Academy of Sciences, 1978. xv + 694 pp.

Includes sections on A. A. Michelson and Frederick Seitz.

C18 Cockcroft, John, Ed. The Organization of Research Establishments. Cambridge: Cambridge University Press, 1965. 275 pp.

Contains: Sir Gordon Sutherland, "The National Physical Laboratory," 6-27; F. A. Vick, "The Atomic Energy Research Establishment, Harwell," 55-77; J.B. Fisk, "The Bell Telephone Laboratories," 197-214; T. G. Pickavance, "The Rutherford High Energy Physics Laboratory," 215-35; J. B. Adams,"CERN: The European Organization for Nuclear Research," 236-61.

C19 Condon, E. U. "The past and future of the Reviews of Modern Physics." Reviews of Modern Physics, 40 (1968): 876-78.

C20 Cooper, John N. "American physicists and their graduate degrees." American Journal of Physics, 20 (1952): 484-87.

Changes in "productivity" of institutions since 1920.

C21 Crew, Henry. "The scientific leadership of the world." Methodist Review, (January 1920): 95-102.

List of 23 achievements in physics in last 300 years, sorted by country.

C22 Davey, Laura Gunn. The History of Women in Physics State College, Pa. 7 pp.

C23 Drew, David E. Science Development: An Evaluation Study. (Techical Report presented to the National Board on Graduate Education) Washington, D.C.: National Academy of Sciences, 1975. xvi + 182 pp.

Effect of the block grants given by the National Science Foundation to selected universities in the 1960s, on size and productivity of science department faculties.

C24 DuBridge, Lee A. "The role of the private foundations in the development and growth of physics and astronomy in the United States." U.S. Philanthropic Foundations: Their History, Structure, Management, and Record, by Warren Weaver, et al. New York: Harper & Row, 1976, Chapter 17.

C25 Elton, Charles F. and Samuel A. Rodgers. "Physics Department Ratings: Another Evaluation." Science, 174 (1971): 565-68.

Institutional ratings of physics departments (e.g. by A. M. Cartter, An Assessment of Quality in Graduate Education, 1966) can be predicted using data in the public domain, such as number of Ph.D.'s awarded in the previous 5 years.

C26 Elton, Charles F. and Sam A. Rodgers. "The Departmental Rating Game: Measure of Quantity or Quality?" Higher Education, 2 (1973): 439-46.

Confirming the results of the previous paper, the authors find that departmentsl reputation in the physical sciences, as assessed by raters in the study of A. M. Cartter, is based primarily on quantitative factors rather than quality.

C27 Feather, Norman. "The Cavendish Laboratory - Aristocrat's playground to people's workshop." Contemporary Physics, 16 (1975): 211-13.

C28 Fitzpatrick, T. C., et al. A History of the Cavendish Laboratory 1871-1910. London: Longmans Green, 1910. x + 342 pp.

Includes articles on the period since 1890 by J. J. Thompson, E. Rutherford, C. T. R. Wilson, and R. N. Campbell.

* Fleury, P. "The International Union of Pure and Applied Physics from 1923 to 1972." Cited above in item A15.

C29 Flowers, Brian. "The Physical Society of London 1874-1974." Physics Bulletin, 25 (1974): 435-37.

C30 Folta, Jaroslav. "Univerzita Karlova a vývoj matematicko-fyzikálnich věd." Vědecká konference matematicko-fyzikální fakulty Univerzity Karlovy. Část I. Matematika. Prague: Univerzita Karlova, 1979, pp. 30-44.

C31 Forman, Paul. "Scientific Internationalism and the Weimar physicists: The Ideology and its Manipulation in Germany after World War I. " Isis, 64 (1973): 151-80.

Forman stresses "the essentially nationalistic foundations and functions of scientific internationalism while emphasizing that the classical formula for the participation of the nation in the scientist's fame spares the scientist any conflict between advancing his science and advancing the interests of his nation...The Wilhelmian Gelehrten endeavored to broaden the classical formula providing national prestige from scientific achievement to an equation between political and scientific great-power status...In the Weimar period, however, following Germany's military and economic prostration, the Gelehrten shifted this rhetoric subtly and significantly to the proposition that scientific great-power status could function as a substitute for political great-power status..."

C32 Forman, Paul. "The Financial Support and Political Alignment of Physicists in Weimar Germany." Minerva, 12 (1974): 39-66.

The blossoming of physical research in the first 10 years of the republican regime "was due in considerable part to the funds provided by the central government...and especially to the distinctive mode of allocating it as grants awarded for specific research projects on the basis of recommendations by panels of highly qualified referees...in post-war Germany, physics had become so deeply and thoroughly affected by 'politics' that every social institution of, or closely associated with, this discipline necessarily also had a more or less strongly marked political character...

C33 Forman, Paul, John L. Heilbron, and Spencer Weart. "Physics _circa_ 1900: Personnel, Funding, and Productivity of the Academic Establishments." _Historical Studies in the Physical Sciences_, 5 (1975): 1-185.

A comprehensive statistical analysis of academic physics in 1900, giving data on numbers of positions, personal income, age of entry into physics posts, laboratory expenditures, grants for research, rate of construction of new facilities, output of physics papers and input of resources devoted to research. On the basis of their "best guess" of the "gross national product of physics papers," the authors conclude that "Germany is far ahead; the British Empire is second, followed by a faltering France; the United States is a distant, but rapidly gaining, fourth; Italy, a poor and declining fifth. The French, unwarrantably complacent about the state of science in their country, and that of physics in particular, neglected to arrest their relative decline, while the Germans, although well aware that they led, bore the challengers ever in mind." (pp. 115,118).

C34 French, A. P. "Fifty years of physics education." _Physics Today_, 34, no. 11 (Nov. 1981): 51-62.

C35 Frenkel, V. Ia. "50 let Sovetskogo Soîuza i sovetskaía fizika." _Uspekhi fizicheskikh Nauk_, 108 (1972): 617-24.

C36 Frenkel V. Ia. "Akademii Nauk SSR 250 let." _Uspekhi Fizicheskikh Nauk_, 113 (1974): 3-28.

C37 Frenkel, V. Ia., "AN SSSR - 250 let." _Fizika Tverdogo Tela_, 16 (1974): 977-80.

C38 Frenkel, V. Ia. "Dvukhsotpîatidesîatiletie Akademii nauk." _Zhurnal Tekhnicheskoi Fiziki_, 44 (1974): 897-905.

C39 Frenkel, V. Ia. "Shestidesiatiletie FTI." _Atomnaia Energiia_, 45 (1978): 389-91.

C40 Frenkel,V. Ia., ed. _Theoretical Physics, Mathematical Physics, Applied Mathematics and Computer Technology_. Leningrad: "Nauka" Publishers, 1978. 64 pp.

Chief Editor Acad. V. M. Tuchkevich, compiled by V. Ia. Frenkel.

"This booklet, printed in connection with the 60th Anniversary of the A.F. Ioffe Physicotechnical Institute of the USSR Academy of Sciences, gives a brief historical survey of research in theoretical physics at the Institute. It presents a synopsis of results obtained in the electron theory of solids (semi-conductors and metals), physical kinetics, applied theoretical physics and crystal lattice physics during the past decade. The second part of the booklet summarizes the research results for the years 1968-1977 in mathematical physics, its general methods and their applications to a number of problems (heat conductivity, elasticity theory, astrophysics, etc.) and in applied mathematics and computer technology.

C41 Frenkel, V. Ia. *A. F. Ioffe Physico-Technical Institute*. USSR Academy of Sciences. Leningrad: "Nauka" Publishers, Leningrad Branch, 1978. 88 pp.

Chief Editor Acad. V. M. Tuchkevich, compiled by V. Ia. Frenkel.

This booklet is the first one of a series printed in connection with the 60th Anniversary of the A.F. Ioffe Physicotechnical Institute of the USSR Academy of Sciences. A short history of the Institute is given. The Institute was named in honor of its founder and first director, the prominent Soviet physicist A. F. Ioffe. The principal part of the book is made up of synopsis of scientific investigations carried out during the last ten years on a wide range of problems: semiconductor and solid state physics, plasma physics, emission and quantum electronics, astrophysics, theoretical and mathematical physics, and others.

C42 Frenkel, V. Ia. "60 let Fiziko-teknicheskomu institu." *Vestnik AN SSSR*, 9 (1979): 64-68; 76-77.

C43 Goodstein, Judith. "Science and Caltech in the Turbulent Thirties." *California History*, 60, no. 3 (1981): 229-243.

C44 Greenberg, Daniel S. *The Politics of Pure Science*. New York: New American Library, 1967. xiii + 303 pp.

ChapterX, "High Energy Politics"; Chapter XI, "MURA's last stand."

C45 Gulo, D. D., A. F. Konokov, A. N. Osinovskiĭ. "Iz Istorii Osnovanii͡a Gosudarstvennogo Opticheskogo Instituta." *Istorii͡a I Metodologii͡a Estestvennykh Nauk*, 3(1965):273-92.

On the occasion of the 45th anniversary of the founding of the State Optical Institute.

C46 Hermann, A. "25 Jahre Physikalische Blaetter." *Physikalische Blaetter*, 25 (1969): 547-551.

C47 Hermann, A. "Forschungsfoerderung der Deutschen Forschungsgemeinschaft und die Physik der letzten 50 Jahre." (Festvortrag anlaesslich des 50-jaehrigen Bestehens der DFGam 30. Oktober 1970 in Berlin. _Deutsche Forschungsgemeinschaft Mitteilungen_, 4/70 (1970): 231-34.

Idem in _Physik in unserer Zeit_, 2 (1971): 17-23.

C48 Hermann, A. "Das goldene Zeitalter der Physik." _Leistung und Schiksal, 300 Jahre juedische Gemeinde in Berlin_. Berlin: Berlin-Museum, 1971, pp. 33-40.

C49 Hermann, A. "Deutsche Wissenschaftspolitik im 20. Jahrhundert." _Umschau aus Wissenschaft und Technik_, 74 (1974): 734-738.

C50 Hermann, A. "130 Jahre Deutsche Physikalische Gesellschaft." Festvortrag. _Physikalische Blaetter_, 31 (1975): 544-47.

C51 Hermann, A. "Deutschlands Weg in das Atomzeitalter." _Bild der Wissenschaft_, 13 (1976): 42-46.

C52 Hermann, A. "Wissenschaftspolitik und Entwicklung der Physik im Deutschen Kaiserreich." _Medizin, Naturwissenschaft, Technik und das Zweite Kaiserreich_. Edited by Gunter Mann and Rolf Winau. Goettingen: Vandenhoeck und Ruprecht, 1977, pp. 52-63.

C53 Hermann, A. "Physik an der Eberhard-Karls-Universitaet." Physik, _Physiologische Chemie und Pharmazie an der Universitaet Tuebingen_. Edited by Armin Hermann and Armin Wankmuller. (Contubernium, Bd. 21). Tübingen: Wolf von Engelhardt, 1980, pp. 13-39.

C54 Horák, Zdeněk. "Pocatky fyzikálních přednášek na univerzitě a technice po vzniku samostatného státu." _Prace z dejiny ved a techniky_, 11 (1979): 57-66.

"The beginnings of the lectures in physics at the Prague universities after the creation of the independent state."

* Jachim, A. G. _Science Policy Making in the United States and the Batavia Accelerator_. Cited below as item JH31.

* Jacob, M., ed. _CERN -- 25 Years of Physics_. Cited below as item JH32.

C55 Janout, Z. "Dvacet pet let FJFI." _Rozhledy matematicko-fyzikalni_, 59 (1980/1981): 34-37.

25 years of the Faculty of Nuclear Science and Physical Engineering, Prague Technical University.

C56 [Kaiser-Wilhelm-Gesellschaft] _50 Jahre Kaiser-Wilhelm-Gesellschaft und Max-Planck-Gesellschaft zur Förderung der Wissenschaften 1911-1961._ Beitraege und Dokumente Herausgegeben von der Generalverwaltung der Max-Planck-Gesellschaft zur Förderung derWissenschaften E.V., Göttingen, 1961. 262 pp.

Contains notes on the Presidents of MPG (Adolf von Harnack, Max Planck, Carl Bosch, Albert Vögler, Otto Hahn, Adolf Butenandt) and a list of documents of MPG history.

C57 Kargon, R. H. "Temple to science: Cooperative research and the birth of the California Institute of Technology." _Historical Studies in the Physical Sciences_, 8 (1977): 3-31.

Focuses on the role of R. A. Millikan, G. E. Hale, and A. A. Noyes in the 1920s. Their aspirations for Caltech "were related to the requirements of research in the borderland between physics, chemistry, and astronomy." They exploited opportunities created by the rapid commercial development of Southern California to obtain resources, cultivating wealthy patrons.

C58 Kelbg, Gunter and Wolf Dietrich Kraeft. "Die Entwicklung der theoretischen Physik in Rostock." _Wissenschaftliche Zeitschrift der Universitaet Rostock. Mathematisch-naturwissenschaftliche Reihe_, 16 (1967): 839-47.

C59 Kevles, Daniel J. _The Physicists: The History of a Scientific Community in Modern America_. New York: Knopf, 1978. xi + 489 pp.

An outstanding, comprehensive treatment of the growth of physics in the United States since the late 19th century, weaving together social, institutional and techical history. Important figures in the story are H. A. Rowland, R. A. Millikan, G. E. Hale, A. Trowbridge, I. I. Rabi, H. C. Urey, C. D. Anderson, K. T. Compton, V. Bush, J. R. Oppenheimer. A. H. Compton, J. B. Wiesner.

C60 Kiger, Joseph C. _American Learned Societies_. Washington, D.C.: Public Affairs Press, 1963. 292 pp.

Includes some information about the American Institute of Physics and the American Physical Society.

C61 Kingslake, Hilda G., et al. "History of the Optical Society of America." _Journal of the Optical Society of America_, 56 (1966): 273-340.

* Kirsten, Christa and Hans-Guenther Koerber, eds. _Physiker ueber Physiker_. Cited below as items B24 & B25.

C62 Kistiakowski, Vera. "Women in physics: Unnecessary, injurious and out of place?" Physics Today, 33, no. 2 (Feb. 1980): 32-40.

C63 Koch, H. William. "AIP Today - and Tomorrow." Physics Today, 34, no. 11 (Nov. 1981): 233-41.

C64 Kol'tsov, A. V. "50-letie zhurnala 'Vestnik Akademii Nauk SSSR'." Voprosy Istorii Estestvoznaniia i Tekhniki, 2 (1981): 130-34.

50 years of the Journal Vestnik Akademii Nauk SSSR.

C65 Konstantinov, B. P. "Razvitie Leningradskoĭ Shkoly Fisiki." Vestnik AN SSSR, 11 (1967): 93-104.

C66 Kowarski, Lew. "Some conclusions from CERN's History." New Phenomena in Subnuclear Physics. Part B. Edited by A. Zichichi. New York: Plenum Press, 1977, pp. 1201-11.

C67 Langevin, Andre. "Paul Langevin et les congres de physique Solvay." La Pensee, 129 (1966): 3-32; 130 (1966): 89-104. Translation: "Pol' Lanzheven i Sol'veevskie Fizicheskie Kongressy." Edited and annotated by O. A. Starosel'skaia-Nikitina. Voprosy Istorii Estestvoznaniîa i Tekhniki, 1 (26), (1969): 3-26.

Paul Langevin and the Solvay Physics Congresses.

C68 Lezhneva, O. A. "Fizika v Akademii Nauk v XIX-Nachale XX Veka." Tezisy Dokladov i Soobschchenii Na Mezhvuzovskoi Konferentsii Po Istorii Fiziko-Matematicheskikh Nauk. Moskva, 1960.

Physics at the Academy of Sciences in the 19th and early 20th centuries.

C69 Libman, E. P. "Russkiĭ Radiĭ." Priroda, 6 (1966): 64-67.

On the history of foundation of the Radium Institute in Petrograd in 1921-22.

C70 Lock. W. O. Origins and evolution of the Collaboration between CERN and the People's Republic of China 1971-1980. Geneva: CERN (European Organization for Nuclear Research), 1981. 49 pp.

C71 Lodahl, Janice Beyer and Gerald Gordon. "The structure of scientific fields and the functioning of university graduate deparetments." American Sociological Review, 37 (1972): 57-72.

A test of Kuhn's paradigm concept applied to physics departments.

C72 Lyons, Henry, Sir. The Royal Society 1660-1940. A History of its Administration under its Charters. Cambridge: Cambridge University Press, 1944. 354 pp.

The last chapter deals with research from 1901 to 1940.

* MacCallum, T. W. and Stephen Taylor, eds. The Nobel Prize-Winners and the Nobel Foundation 1901-1937. Cited above as item B31.

C73 Mehra, Jagdish. The Solvay Conferences on Physics. Aspects of the Development of Physics since 1911. Dordrecht/Boston: D. Reidel, 1975. xxxii + 415 pp. With a forward by Werner Heisenberg, pp. v-vii.

Includes detailed synopses of the papers and discussions at each meeting from 1911 to 1973.

C74 Michaelis, Anthony. "The Recovery of Science in Germany." Interdisciplinary Science Reviews, 6, no. 4 (December 1981): 283-311.

The author, a German-born refugee in Britain since 1933, followsthe reconstruction of German science since the end of World War 2 and argues that the general level of science in Germany has yet to equal the heights of the thirties andthat there was a major recovery of science in Germany butnot a complete one.

C75 Micheli, Gianni, ed. Storia d'Italia, Annali 3, Scienza e tecnica nella cultura e nella societa dal Rinascimento a oggi. Torino: Einaudi, 1980. 1365 pp.

Includes: Roberto Maiocchi , "Il ruolo delle scienze nello sviluppo industriale italiano," 865-999. The author tries to estabish a link between scientific and industrial developments.

C76 Mikulinskij, S. R., M. G. Jarosevskij, G. Krober, H. Steiner, eds. Wissenschaftliche Schulen. Berlin: Akademie Verlag, 1977-79. Bd. 1, 449 pp.; Bd. 2, 208 pp.

C77 Mikulinskii, S. R., M. G. Iaroshevskii, G. Krober, G. Shteiner, Eds. Shkoly v nauke. Moskva: Nauka, 1977. 523 pp.

C78 Miroshnikov, M. M. Ed. 50 Let Gosudarstvennogo Opticheskogo Instituta im. S. I. Vavilova. 1918-1968. Leningrad: Mashinostroenie, 1968. 708 pp.

Essays on the history of State Optical Institute S. I. Vavilov from 1918 to 1968.

C79 Moffat, Linda K. "Departmental characteristics and Physics Ph.D. production 1968-1973." Sociology of Education, 51 (1968): 124-32.

C80 Moseley, Russell. "Government science and the Royal Society: The control of the National Physical Laboratory in the inter-war years." Notes and Records of the Royal Society of London, 35 (1980): 167-93.

C81 Mulkay, Michael J. and Anthony T. Williams. "A Sociological study of a physics department." British Journal of Sociology, 22 (1971): 68-82.

Publications and reward system in a Canadian university department.

C82 Nenicka, Miloslav. "Prameny k dejinam fysikalnich a technickych ustavu CSAV uchovavane v UA CSAV. I. cast (Ustav technicke fyziky CSAV, Ustav pro elektrotechniku CSAV, Hutnicky ustav CSAV." Archivní zprávy Československá Akademie Věd. 7 (1975): 119-27.

Sources for the history of physical and technical institutes of the Czechoslovak Academy of Sciences (CSAS) in the Central Archive of CSAS, Part I. For Part II see ibid. 9 (1977): 120-34; for Part III see ibid. 11/12 (1979-80): 111-27.

C83 Nobel Foundation,ed. Nobel: The Man and His Prizes. Second Edition. New York: American Elsevier, 1962. x + 690 pp.

Includes Manne Siegbahn, "The Physics Price," 439-518.

C84 Nobel Foundation. Nobel: The Man and His Prizes. Third Edition. New York: American Elsevier, 1972. x + 659 pp. Edited by the Nobel Foundation and W. Odelberg, Coordinating Editor.

Includes: Manne and Kai Siegbahn, "The Physics Prize," 387-475.

C85 Owens, A. J. "Will zero-population growth hamper scientific creativity?" Physics Today, 26, no. 10 (October 1973): 9, 11.

Includes age distribution of Nobel-Prize-winning work in physics.

C86 Parkadze, V. D. and M. V. Parkadze. "On the Scientific Connections Between Physicists of Great Britain and Georgia (1700-1975)." Papers by Soviet Scientists. (item M48), pp. 56-61.

C87 Piore, Emanuel R. "Physics funding--the evolving federal role." Physics Today, 34, no. 12 (Dec. 1981): 38-45.

History of government support for physics in U. S. in last 50 years.

C88 Przibram, K. "Erinnerungen an eines altes physikalisches Institut." Beitrage zur Physik und Chemie des 20. Jahrhunderts (item B11) Vieweg: Braunschweig, 1959, pp. 1-6.

On Vienna physics, circa 1900.

C89 Pyenson, Lewis, and Douglas Skopp. "Educating physicists in Germany circa 1900." Social Studies of Science, 7 (1977): 329-66.

The authors analyze the social and educational backgrounds of Ph.D. recipients in the philosophical faculties at German universities in the decades before the First World War. "Some expected patterns and several striking contrasts emerge from a study of student origins and migrant patterns from university to university, as well as of those professors who exerted great influence in determining the institutional configurations of physical sciences..." Contrary to Fritz Ringer's thesis (Decline of the German Mandarins, 1969) that "new financial and entrepreneurial groups did not send their sons in large numbers to the universities," they find that "around 1900, the bourgeoisie and petty bourgeoisie contributed heavily to the Ph.D. rolls." Many of them went on to hold mandarin positions in the late 1920s and might have offered little or no resistance to fascism.

C90 Pyenson, Lewis. "Mathematics, education, and the Goettingen approach to physical reality, 1890-1914." _Europa_, 2 (1979): 91-126.

Presents an example of how pressures "from below" (secondary mathematics schoolteachers) interacted with the interests and prejudices of those at the top (university professors of physics). Felix Klein played a central role in the reform of mathematics teaching, and in injecting instruction inpure mathematics into the training of scientists and engineers, epecially at Goettingen. One result was the works of David Hilbert and Gustav Mie in relativity theory, and more generally an ideology of "mathematical instrumentalism" in approaching problems in physics.

C91 Querner, Hans and Heinrich Schipperges, Eds. _Wege der Naturforschung, 1822-1972, im Spiegel der Versammlungen Deutscher Naturforscher und Arzte_. Berlin: Springer, 1972. vi + 207 pp.

Includes: Ulrich Benz, "Quantum- und Relativitaetstheorie im Spiegel der Naturforscherversammlungen, 1906-1920."

C92 Ratcliffe, J. A. "The history of the Physical Society." _Physics Bulletin_, 25 (1974): 355-58.

On the founding of the Physical Society of London in 1874 with brief remarks on its later development.

C93 Reif, F. "The competitive world of the pure scientist." _Science_, 134 (1961): 1957-62.

A personal report (frequently cited by later writers on the sociology of science) apparently inspired by the founding of _Physical Review Letters_.

C94 Reingold, Nathan. "The case of the disappearing laboratory." _American Quarterly_, 29 (1977): 80-100.

On R. A. Millikan, the National Research Council, Rockefeller Foundation, etc.

C95 Richter, Steffen. "Die Kämpfe innerhalb der Physik in Deutschland nach dem ersten Weltkrieg." _Sudhoffs Archiv_, 57 (1973): 195-207.

On the political activities of J. Stark and others in Germany after World War I.

C96 Richter, Steffen. "Wirtschaft und Forschung." _Technikgeschichte_, 46 (1979): 20-45.

C97 Roller, Duane. "The periodical literature of physics: some of its history, characteristics and trends." _American Journal of Physics_, 14 (1946): 300-8.

C98 [Rozhdestvenskiĭ, D. S.] Osinovskiĭ, A. N. "D. S. Rozhdestvenskiĭ - Borets za Nerazrryvnuîu Sviaz' Nauki s Proizvodstvom." _Voprosy Istorii Fiziki i ee Prepodavaniîa_. Tambov, 1961, pp. 81-89.

On D. S. Rozhdestvenskiĭ's views on the connection between science and production.

C99 Snow, C. P. _Science and Government_. Cambridge, Mass.: Harvard University Press, 1961. 88 pp. With an appendix published separately, 1963, 37 pp.

On H. Tizard and F. Lindemann, Lord Cherwell, as advisors to the British government in World War II.

C100 Spasskiĭ, B. I., L. V. Levshin, and V. A. Krasil'nikov. "Physics and astronomy at Moscow University (on the University's 225th anniversary)." _Soviet Physics Uspekhi_, 23 (1980): 78-93. Translated from _Uspekhi Fizicheskikh Nauk_, 130 (1980): 149-75.

C101 Sredniaway, Bronislaw. "The anthropologist as a young physicist: Bronislaw Malinowski's apprenticeship." _Isis_, 72 (1981): 613-20.

On the teaching of physics at the Jagellonian University of Cracow, Poland, 1902-4; remarks on August Witkowski and Wladyslaw Natanson.

C102 Starosel'skaîa-Nikitina, O. A. "Iz Istorii Mezhdunarodnykh Sol'veevskikh Fizicheskikh Kongressov." _Trudy Instituta Istorii Estestvoznaniîa i Tekhniki AN SSSR_, 34 (1960): 9-63.

On seven Solvay Congresses between 1900 and 1933.

C103 Tagliagambe, Silvano. _Scienza, filosofia, politica in Unione Sovietica 1924-1939_. Milano: Feltrinelli, 1978. 524 pp.

Contains articles from journals such as _Pod znamenem marksizma_ and _Estestvoznaniia i marksizm_, largely unknown in the West, on issues of science policy, methodology of science and marxism, etc.

C104 Tesinska, Emilie. "Disertacni prace z fyziky na prazske univerzite v letech 1918-1939." Dejiny ved a techniky, 11 (1978): 113-22.

"Doctoral dissertations in physics at the University of Prague from 1918 to 1939."

C105 Trlifaj, Miroslav. "Vyvoj fyzikalnich pracovist v CSAV." Prace z dejin prirodnich ved, 4 (1973): 251-56.

Development of physical establishments in the Czechoslovak Academy of Sciences.

C106 Tuchkevich, V. M. "Kolybel' Sovetskoi Fiziki." Priroda, 6 (1969): 3-12.

The cradle of Soviet Physics: Physico-Technical Institute, Leningrad.

C107 Tuchkevich, V.. M. "Kuznitsa Sovetskoi Viziki." Sovremennye Fizicheskie Issledovaniia. Moskva: Zhanie, 1969: 3-10.

Short history of the Physico-Technical Institute.

C108 Ulehla, Ivan. "O vedecke cinnosti ve fyzice na vysokych skolach v ceskych zemich v letech 1945-1975." Matematika a fyzika ve skole, 5 (1974-75): 600-8.

"On the scientific activities in physics at universities in Czech lands in the years 1945-1975."

C109 Vacek, K. "The Twenty-Fifth Anniversary of the College of Mathematics and Physics of Charles University." Acta Universities Carolinae, Mathematica et Physica, 19 (1978): c. 2, pp. 5-6.

C110 Van Hove, L. and M. Jacob. "Highlights of 25 Years of Physics at CERN." Physics Reports, 62 (1980): 1-86.

C111 Vinokurov, B. Z. "Kharakteristika Sostoianiia Russkoi Fiziki v Nachale XX Veka." Voprosy Istorii Fiziki i ee Prepodavaniia. Tambov, 1961, pp. 165-178.

On physics education, physics journals and scientific societies in Russia in early 20th century.

C112 Visher, Stephen Sargent. Scientists Starred 1903-1943 in American Men of Science: A Study of Collegiate and Doctoral Training, Birthplace, Distribution, Backgrounds, and Developmental Influences. Baltimore: Johns Hopkins Press, 1947. xxiii + 556 pp.

C113 Vlachy, Jan. "Czechoslovak Academy of Sciences and its physics institutes, 1952-1972." Czechoslovak Journal of Physics, sekce B, 22 (1972): 1202-9.

C114 Weeks, Dorothy W. "Women in physics today." Physics Today, 13, no. 8 (Aug. 1960): 22-23.

Mentions some women who have been honored, and gives numbers of degrees granted in physics, 1949-53.

C115 Weinberg, Alvin M. Reflections on Big Science. Cambridge, Mass.: MIT Press, 1967. ix + 182 pp.

Includes his influential 1963 article, "Criteria for scientific choice."

C116 Weiner, Charles. " A new site for the seminar: The refugees and American physics in the thirties." Perspectives in American History, 2 (1968): 190-234.

Discusses the interactions of American and European physics in the 1920s, the dissolution of German physics in the early 1930s, and attempts of the international scientific communities to aid displaced scientists. The experiences of V. F. Weisskopf and H. A. Bethe are typical of successful absorption into American physics departments. Includes a letter from R. Ladenburg and E. Wigner in 1933 to 27 American physicists, asking for assistance.

* Weiner, Charles. "Institutional settings for scientific change: episodes from the history of nuclear physics." Cited below as item JF70.

C117 Whitley, Richard. "Changes in the social and intellectual organization of the sciences: professionalization and the Arithmetic Ideal." The Social Production of Scientific Knowledge. Edited by E. Mendelsohn et al. Boston: Reidel, 1977, pp. 143-69.

Includes comments on why physics has become the dominant science and why theoretical physicists dominate experimentalists.

C118 Wigner, E. P., et al. Physical Science and Human Values: A Symposium. Princeton, N.J.: Princeton University Press, 1947. vii + 181 pp.

Includes papers by I. I. Rabi, L. A. DuBridge, and P. W. Bridgman.

C119 Zuckerman, Harriet and Robert K. Merton. "Sociology of refereeing." Physics Today, 24, no. 7 (July 1971): 28-33.

On papers submitted to Physical Review. See also letters to the editor, ibid. 25, no. 2 (Feb. 1972): 9, 11.

C120 Zuckerman, Harriet and Robert K. Merton. "Patterns of evaluation in science: Institutionalisation, structure and functions of the referee system." Minerva, 9 (1971): 66-100.

Analysis of The Physical Review and comparison of journals in different fields.

* * * * * * * * * * * * *

For additional references see HW, sections C, A, and U.

D. MECHANICS

D1 Fradlin, B. N. and V. M. Sliusarenko. "O nekotory issledovaniiakh otechestvennykh uchenykh po dinamiko tverdogo tela." *Voprosy Istorii Estestvoznaniia i Tekhniki*, 54, no. 1 (1976): 46-48.

On some investigations by Russian scientists concerning solid state dynamics. Mentions the contributions of A. M. Liapunov, S. A. Chaplygin, V. N. Steklov and others.

D2 Goldstein, Herbert. "This Week's Citation Classic: Goldstein, H. *Classical Mechanics*, Cambridge, MA: Addison-Wesley, 1950. 399 p. *Current Contents, Engineering, Technology & Aplied Sciences*, 12, no. 2 (January 12, 1981): 16.

The author recalls the circumstances leading to the writing of this book, which has been cited over 1,435 times since 1961.

D3 Grigor'ian, A. T. "Die Mechanik der Körper veränderlicher Masse und ihre Entwicklung in der USSR." *MTM-Schriftenreihe für Geschichte der Naturwissenschaften, Technik und Medizin*. 11, no. 2 (1974): 44-57.

On the work of I. V. Mescerskij (1859-1935) and K. E. Ciolkovskij (Tsiolkovsky) (1857-1935) published in the period 1897-1929, and subsequent applications to the dynamics of rocket flight by other Russian scientists and engineers.

D4 Grigor'îan, A. T. "Istoriia Mekhaniki." *Voprosy Istorii Estestvoznaniia i Tekhniki*, 59, no. 2 (1977): 42-46.

On research in the history of mechanics in the USSR during the last 50 years.

D5 Grigor'ian Ashot Tigranovich. Ocherk Razvitiia Mekhaniki v SSSR. Moskva: Izdatel'stvo "Russkii Iazyk," 1979. 277 pp.

The first chapter deals with the development of mechanics in Russia in the 18th century, while the second treats the period 19th century to beginning of 20th century. Chapter 3 highlights directions of developments of mechanics in USSR. Chapter 4 contains biographies of M. V. Keldish, S. P. Korolev, A. N. Krylov, A. N. Tupolëv, A. A. Fridman, K. E. Tsiolkovskiĭ, S. A. Chaplygin, N. G. Chetaev and others. The book ends with an index of mechanical terms with their English, German and French translations. The book is intended for foreign students wishing to study mechanical engineering in Soviet Polytechnics, providing information on the background of different institutions, their specializations (theory of vibrations, hydro-aerodynamics) and trends. Intends to familiarize students also with Soviet scientific style and terminology. The book is of interest to both linguists and historians of science.

D6 Grigor'ian. A. T. "The chief stages in the evolution of mechanics in Russia." Janus, 68 (1981): 27-32.

D7 Odqvist, Folke K. G. "Nonlinear mechanics, past, present, and future." Applied Mechanics Reviews, 21 (1968): 1213-22.

On the work of G. Truesdell and others.

D8 Signorini, Antonio. "Contributions Italiennes a la Mecanique Theorique de Leonard de Vinci a Levi-Civita." Cahiers d'Historie Mondiale, 7 (1963): 419-33.

Inludes brief remarks on the work of V. Volterra, C. Somigliana, E. Almansi, G. A. Maggi, R. Marcolongo, G. Giorgi, and a longer section on T. Levi-Civita.

D9 Truesdell, G. "Reactions of the History of Mechanics upon modern research." Essays in the History of Mechanics. New York: Springer-Verlag, 1968, 305-33. Reprinted from Proceedings of the Fourth U.S. National Congress of Applied Mechanics (Berkeley, 1962).

"It is shown by examples that knowledge of the history of mechanics can lead to new discoveries in the mechanics of today. The examples adduced concern: (a) the strongest form of column for axial thrust. (b) the form to be given a beam in order that when loaded it shall take on a desired shape. (c) Invariance requirements for the constitutive equations of non-linear materials." (Author's abstract).

D10 Truesdell, C. "History of classical mechanics." *Naturwissenschaften*, 63 (1976): 53-62, 119-30.

The second part covers the 19th and 20th centuries.

* * * *

For additional references on this topic see HW, Section D.

DA. PROPERTIES OF BULK MATTER: THERMODYNAMICS

DA1 Bridgman, Percy W. "General survey of certain results in the field of high-pressure physics." Nobel Lectures...Physics (item A87), vol. 3, pp. 53-70. December 11, 1946.

DA2 Carathéodory, C. "Investigation into the foundations of thermodynamics." Translated from Mathematische Annalen, 1909. The Second Law of Thermodynamics. Edited by Joseph Kestin. Stroudsburg, PA: Dowden, Hutchinson & Ross, 1976, pp. 229-56.

DA3 Drost-Hansen, W. "This Week's Citation Classic: Drost-Hansen W. Structure of water near solid interfaces. Ind. Eng. Chem. 61(11): 10-47, 1969." Current Contents, Physical, Chemical & Earth Sciences, 20, no. 10 (March 10, 1980): 12.

The author recalls the circumstances of writing this paper, which has been cited more than 125 times since 1969.

DA4 Frankfurt, U. I. "K istorii aksiomatiki termodinamiki." Razvitie Sovremennoi Fiziki (item A62), pp. 257-92.

On the history of axiomatic thermodynamics.

DA5 Gopal, E. S. R. and B. Viswanathan. "One hundred years of critical point phenomena." Journal of Scientific and Industrial Research (New Delhi), 28 (1969): 204-14.

DA6 Kapustinskii, A. F. "Maks Plank kak termodinamik i fiziko-khimik." Maks Plank. 1858-1958 (item BP17), pp.109-25.

Max Planck as thermodynamicist and physico-chemist.

DA7 Kelker, H. "History of Liquid Crystals." Molecular Crystals and Liquid Crystals, 21, no. 1-2 (1973): 1-48.

DA8 Kubo, Ryogo. "Statistical Mechanics: A Survey of its One Hundred Years." Scientific Culture (item A78), pp. 131-57.

Broad semi-historical survey from Boltzmann to Onsager and H. Haken.

DA9 Kurti, N. "From Cailletet and Pictet to Microkelvin." Cryogenics, 18 (1978): 451-8.

Details on the history of air liquefaction.

DA10 Kuznetzova, O. V. "Razvitie teorii agregatnykh sostoianii veshchestva." Voprosy Istorii Estestvoznaniia i Tekhniki, no. 3-4 (56-57) (1977): 66-68.

Development of the theory of states of aggregation of matter based on the interaction of particles and the statistical principles of J. W. Gibbs. Also considers the interpretation of superfluidity and superconductivity.

DA11 Kvasnica, J. "Nerelavisticke dilo Alberta Einsteina." Casopsis pro Pestovani Fysiky., Sekce A, 29 (1979): 212-23.

Summarizes Einstein's work on Brownian motion, photo-effect, statistical mechanics, solid state.

DA12 Levelt Sengers, J. H. M. "Critical exponents at the turn of the century." Physica, 82a (1976): 319-51.

DA13 Levelt Sengers, J. H. M. "Liquidons and Gasons: Controversies about the Continuity of States." Physica, 98A (1979): 363-402.

DA14 Piech, Tadeusz. Wklad Polskich Uczonych do Fizyki Statystyczno-Molekularnej. Wroclaw: Zaklad Narodowy Imienia Ossolińskich Wydawnictwo Polskiej Akademie Nauk, 1962. xxv + 279 pp.

Includes reprints (in Polish) of papers by L. Natanson, M.v. Smoluchowski, etc.

DA15 Predvoditelev, A. S. "O problemakh, sviazannykh s prevrashcheniem veshchestva v gomogennykh i geterogennykh sistemakh." Razvitie Fiziki v Rossii (item A93), pp. 155-76.

On the change of matterfrom homogeneousto heterogeneous systems.

DA16 Rayleigh, Lord (John William Strutt). "The density of gases in the air and the discovery of argon." Nobel Lectures...Physics (item A87), vol. I, pp. 90-96. December 12, 1904.

DA17 Shimony, Abner. "Carnap on Entropy. Introduction to 'Two Essays on Entropy' by Rudolf Carnap." Rudolf Carnap, Logical Empiricist. Edited by Jaakko Hintikka. Dordrecht: Reidel Pub. Co., 1975, pp. 381-95.

* * * * *

For additional references see HW, section R.d

DB. STATISTICAL MECHANICS

DB1 Baracca, Angelo. _Manuale critico di meccanica statistica_. Catania: CULC, 1980. 661 pp.

Graduate textbook in statistical mechanics with much historical and critical apparatus.

DB2 Barker, J. A. "Gibbs' contribution to statistical mechanics." _Proceedings of the Royal Australian Chemical Institute_, May 1976, pp. 131-37.

DB3 Barker, J. A. and D. Henderson. "This Week's Citation Classic: Perturbation theory and equation of state for fluids. II. A successful theory of liquids. _J. Chem. Phys._ 47: 4714-21, 1967." _Current Contents, Engineering, Technology & Applied Sciences_, 12, no. 7 (February 16, 1981): 16.

The author recalls the circumstances of writing this paper, which has been cited more than 300 times since 1967.

DB4 Bonner, Jill C. "This Week's Citation Classic: Bonner, J.C. & M. E. Fisher. Linear magnetic chains with anisotropic coupling. _Phys. Rev._ A 135: 640-58, 1964." _Current Contents, Physical, Chemical & Earth Sciences_ 20, no. 38 (22 Sept. 1980): 16.

Bonner recalls the preparation and reception of this paper, which has been cited over 355 times since 1964.

DB5 Brush, Stephen G. "Functional integrals and statistical physics." _Reviews of Modern Physics_, 33 (1961): 79-92.

On the path-integral version of quantum mechanics proposed by R.P. Feynman, its relation to N. Wiener's theory of Brownian movement, and its application to statistical mechanics by Feynman, R. Kikuchi, R. Abe, and A. M. Yaglom.

DB6 Brush, Stephen G. "History of the Lenz-Ising model." _Reviews of Modern Physics_. 39 (1967): 883-93.

Many physico-chemical systems can be represented more or less accurately by a lattice arrangement of molecules with nearest-neighbor interactions. The simplest and most popular version of this theory is the so-called "Ising model," discussed by Ernst Ising in 1925 but suggested earlier (1920) by Wilhelm Lenz. The paper reviews early approximate methods of solution, the exact solution of the two-dimensional case by L. Onsager, and the mathematically equivalent "lattice-gas" model for gas-liquid and liquid-solid transitions.

DB7 Brush, Stephen G. "Foundations of statistical mechanics 1845-1915." _Archive for History of Exact Sciences_, 4 (1967): 145-83.

The "ergodic hypothesis," attributed to L. Boltzmann and J.C. Maxwell, was first formulated in mathematical terms by P. and T. Ehrenfest. This led to the proof by M. Plancherel and A. Rosenthal (based on mathematical results of G. Cantor, H. Lebesgue, and L. E. J. Brouwer) that mechanical systems can not be ergodic in the strict sense.

DB8 Brush, Stephen G. "Interatomic forces and gas theory from Newton to Lennard-Jones." _Archive for Rational Mechanics and Analysis_, 39 (1970): 1-29.

Includes the development of the force law proposed by J. E. Lennard Jones, a sum of inverse-power attraction and repulsion terms, and subsequent criticism of the notion that this is the most "realistic" formula for interatomic forces.

DB9 Brush, Stephen G. "Proof of the impossibility of ergodic systems: The 1913 papers of Rosenthal and Plancherel." _Transport Theory and Statistical Physics_, 1 (1971): 287-311.

Includes translations of the papers by Arthur Rosenthal and Michel Plancherel.

DB10 Brush, Stephen G. "Statistical mechanics and the philosophy of science." _PSA 1976_ (Proceedings of the Philosophy of Science Association meeting at Chicago, October 1976). Edited by F. Suppe and P. D. Asquith. East Lansing, Mich.: Philosophy of Science Association, 1977, pp. 551-84.

DB11 Byrne, Patrick H. "Statistical and causal concepts in Einstein's early thought." _Annals of Science_, 37 (1980): 215-28.

Einstein thought statistical laws are based on non-statistical assumptions; his later opposition to quantum mechanics was because of, not in spite of, his earliest work in statistical physics.

DB12 De Boer, J. "Van der Waals in his time and the present revival." _Physica_, 73 (1974): 1-27.

DB13 Delbruck, M. "Was Bose-Einstein statistics arrived at by serendipity?" _Journal of Chemical Education_, 57 (1980): 467-70.

DB14 Domb, C. "The Curie Point." Statistical Mechanics at the Turn of the Decade. Edited by E. G. D. Cohen. New York: Dekker, 1971, pp. 81-128.

Semi-historical survey of recent progress in the theory of phase transitions.

DB15 Domb, Cyril. "The solution of some intractable problems." Physikalische Blaetter, 37 (1981): 167-71.

Survey of various problems, mostly relating to phase transitions.

DB17 Dugas, Rene. "Einstein et Gibbs devant la thermodynamique statistique." Comptes Rendus hebdomadaires des Seances de l'Academie des Sciences, Paris, 241 (1955): 1685-87.

DB18 Ezawa, Hiroshi. "Einstein's Contribution to Statistical Mechanics." Albert Einstein: His Influence on Physics, Philosophy and Politics (item BE8), pp. 69-87.

DB19 Ezawa, Hiroshi. "Einstein's contribution to statistical mechanics, classical and quantum." Japanese Studies in History of Science, 18 (1979): 28-72.

DB20 Fierz, Markus. "Statistische Mechanik." Theoretical Physics in the Twentieth Century (item BP4), pp.161-86.

DB21 Fisher, Michael E. "This week's citation classic. 1 Fisher, M. E. The Theory of equilibrium critical phenomena. Rep. Progr. Phys 30: 615-730, 1967." Current Contents, Physical, Chemical & Earth Sciences, 20, no. 46 (Nov., 17 1980): 18.

The author recalls the circumstances of writing this article and his problems with the editors of the journal. The article has been cited over 935 times since 1967.

DB22 Fowkes, Frederick M. "This Week's Citation Classic: Fowkes, F. M. Attractive forces at interfaces. Ind. Eng. Chem. 56: 40-52, 1964." Current Contents, Physical, Chemical & Earth Sciences, 20, no. 18 (May 5, 1980): 12.

The author recalls the circumstances of writing this paper which has been cited more than 195 times since 1964.

DB23 Haas, Arthur, Ed. _A Commentary on the Scientific Writings of J. Willard Gibbs_, Vol. II, Theoretical Physics. New Haven: Yale University Press, 1936. Reprinted, New York: Arno Press, 1980.

Includes articles by Haas and P. S. Epstein on Gibbs' work in statistical mechanics.

DB24 Hildebrand, J. H. "A History of Solution Theory." _Annual Reviews of Physical Chemistry_, 32 (1981): 1-23.

Includes a critique of lattice/cell models.

DB25 Kac, M. "The work of T. H. Berlin in statistical mechanics:..a personal reminiscence." _Physics Today_, 17, no. 10 (October 1964): 40-42.

DB26 Kuznetsova, O. V. and Iu. A. Kukharenko. "Al'bert Eĭnshteĭn i Problema Obosnovaniîa Statisticheskoĕ Mekhaniki." _Eĭnshteĭn i Razvitie Fiziki_. Edited by G. M. Idlis, Moskva: 1981.

Albert Einstein and the problem of foundation of statistical mechanics.

DB27 Landsberg, P. T. "Einstein and Statistical Thermodynamics." _European Journal of Physics_, 2 (1981): 203-7, 208-12, 213-19.

The influence of Einstein on statistical thermodynamics is illustrated by considering three topics from both the historical and a modern point of view: the relativistic transformation of thermodynamic quantities..., the quantisation of the energy of the harmonic oscillator and how it connects with the theory of black-body radiation, and the diffusion-mobility relation for gases and for electrons and holes in solids." (Author's abstract)

DB28 Montroll, Elliott. "One Hundred Years of Statistical Mechanics." _The Many-Body Problem_. Edited by J. K. Percus. New York: Interscience, 1973, pp. 525-33.

DB29 Pihl, Mogens. "Den statistike termodynamiks grundlag i elementaer fremstillung." _Fesstskrift til Niels Bohr_ (item BB9), pp. 101-211.

Discusses Bohr's views on statistical thermodynamics.

DB30 Rossis, George. "Sur la discussion par Brillouin de l'éxperience de pensée de Maxwell." Fundamenta Scientiae 2 (1981): 37-44.

Critique of L. Brillouin's treatment of "Maxwell's Demon."

DB31 Rossis, George. "La logique des expériences de pensée et l'éxperience de Szilard." Fundamenta Scientiae, 2 (1981): 151-62.

Popper and Feyerabend have refuted the validity of Szilard's thought experiment; it does not demonstrate the necessary of the identity between negentropy and information.

DB32 Schwartz, J. "The pernicious influence of mathematics on science." Logic, Methodology, and Philosophy of Science, Edited by E. Nagel et al. Stanford, Calif.: Stanford University Press, 1962, pp. 356-360.

Remarks on Birkhoff's ergodic theorem.

DB33 Semenchenko, V. K. "J. W. Gibbs and his basic work on thermodynamics and statistical mechanics." Uspekhi Khimii, 22 (1953): 1278-84.

DB34 Uhlenbeck, G. E. "Some Historical and critical remarks about the theory of phase-transitions." The Ta-You Festschrift:Science of Matter. Edited by S. Fujita. London: Gordon and Breach, 1978, pp. 99-107.

"These notes are intended as a very modest beginning of a historical and critical presentation of the theory of phase-transitions in the manner of Mach."

DB35 Waals, Van der, Johannes D. "The equation of state for gases and liquids." Nobel Lectures..Physics (item A87), Vol. I, pp. 254-65. December 12, 1910.

* Zubarev, D. N. "This Week's Citation Classic: Zubarev D. N. Double-time Green functions in statistical physics..." Cited below as item DC11.

* * * *

For additional references see HW, section D.h

DC. KINETIC THEORY (TRANSPORT PROPERTIES)

DC1 Brush, Stephen G. "Development of the Kinetic Theory of Gases. VI. Viscosity." American Journal of Physics, 30 (1962): 269-81

Includes a summary of the work of S. Chapman (1916) and D. Enskog (1917, 1922).

DC2 Brush, Stephen G. "Theories of Liquid Viscosity." Chemical Reviews, 62 (1962): 513-48.

Semi-historical survey of research on viscous flow theory, rheology, turbulence, microscopic theories of dense gases and liquids, fluctuation-dissipation theory, quantum-mechanical generalizations, and quantum hydrodynamics (theory of L. D. Landau and I. M. Khalatnikov, 1949). 584 references.

DC3 Brush, Stephen G. Kinetic Theory. Volume 3. The Chapman-Enskog Solution of the Transport Equation for Moderately Dense Gases. Oxford and New York: Pergamon Press, 1972. x + 283 pp.

Contents: The work of D. Hilbert, S. Chapman, and D. Enskog on solutions of the Maxwell-Boltzmann transport equations; determination of intermolecular forces; propagation of sound in monatomic gases; alternatives to the Chapman-Enksog method and mathematical problems (work of T. Carleman, H. Grad, E. Wild, D. Morgenstern, E. Ikenberry, C. Truesdell); generalizations to higher density (G. Jaeger, J. Yvon, M. Born, H. S. Green, J. G. Kirkwood, N. N. Bogoliubov, S. T. Choh). Includes translations of papers by Hilbert (1912) and Enskog (1917, 1922) and reprints of papers by S. Chapman (1916, 1917, 1966); Chapman and F. W. Dootson on thermal diffusion (1917).

DC4 Brush, Stephen G. The Kind of Motion we Call Heat: A History of the Kinetic Theory of Gases in the 19th Century. New York: American Elsevier Pub. Co., 1976. xxxix + 769 pp.

Includes items DB7, DC1, and DD2.

DC5 Chapman, Sydney. "The Kinetic Theory of Gases Fifty Years Ago." Lectures in Theoretical Physics, Volume IX-C, Kinetic Theory. New York: Gordon & Breach, 1967, pp. 1-13.

DC6 Cohen, Morrel H. "This week's citation classic: Cohen, M. H. & Turnbull, D. Molecular transport in liquids and glasses. J. Chem. Phys. 31: 1164-9, 1959." Current Contents, Physical, Chemical & Earth Sciences, 20, no. 43 (27 Oct. 1980): 10.

Cohen recalls the circumstances of writing the paper and its history after publication; it has been cited over 540 times since 1961.

DC7 Monchick, Louis. "This Week's Citation Classic: Monchick L & E. A. Mason. Transport properties of polar gases. J. Chem. Phys. 35: 1676-97, 1961." Current Contents, Engineering, Technology & Applied Sciences, 12, no. 4 (January 26, 1981): 16.

The author recalls the circumstances of writing this paper, which has been cited more than 280 times since 1961.

* Mott, Nevill. "This Weeks' Citation Classic: Mott, N. F. Electrons in disordered structures..." Cited below as item DG55.

DC8 Mountain, Raymond D. "This week's citation classic: Mountain, R. D. Spectral distribution of scattered light in a simple fluid. Rev. Mod. Phys. 38: 205-14, 1966." Current Contents, Physical, Chemical & Earth Sciences, 20, no. 48 (1 Dec. 1980): 10.

The author recalls the circumstances under which he wrote this paper, which has been cited more than 250 times since 1966.

DC9 Prigogine, I., and P. Glansdorff. "L'Ecart a l'equilibre interprete comme une source d'ordre. Structure dissipatives." Bulletin de la Classe des Sciences, Academie Royale de Belgique, 59 (1973): 672-702.

Review of research on nonequilibrium thermodynamics by T. De Donder and others in the "Brussels school."

DC10 Schmeidler, Werner. "Begründung der kinetischen Gastheorie durch D. Hilbert." Integralgleichungen mit Anwendungen in Physik und Technik. I. Lineare Integralgleichungen. Leipzig: Akademische Verlagsgesellschaft Geest & Portig K.-G., 1950, pp. 364-77.

DC11 Zubarev, D. N. "This Week's Citation Classic: Zubarev, D. N. Double-time Green functions in statistical physics. Usp. Fiiz. Nauk. SSSR 71: 71-116, 1960." Current Contents, Physical, Chemical & Earth Sciences, 21, no. 23 (June 8, 1981): 14.

The author recalls the circumstances of writing this paper, which has been cited more than 875 times since 1961.

DD. BROWNIAN MOVEMENT AND FLUCTUATIONS

DD1 Bent, Henry A. "Einstein and chemical thought. Atomism extended." *Journal of Chemical Education*, 57 (1980): 395-405.

Centennial appreciation of his work on Brownian movement, quantum theory and radiation.

DD2 Brush, Stephen G. "A History of Random Processes. I. Brownian movement from Brown to Perrin." *Archive for History of Exact Sciences*, 5 (1968): 1-36. Reprinted in *Studies in the History of Statistics and Probability*, II. Edited by M. Kendall and R. L. Plackett. New York: Macmillan, 1977, pp. 347-82. Also reprinted in item DC4.

Includes an account of the theories of Einstein and M. v. Smoluchowski, and the experiments of J. Perrin which helped to establish the "real existence" of the atom.

* Carazza, B. "The History of the Random-Walk Problem: Considerations on interdisciplinarity in modern physics." Cited above as item A19.

DD3 Einstein, Albert. *Untersuchungen über die Theorie der Brownschen Bewegungen*. Edited by R. Furth. Leipzig: Akademische Verlagsgeselschaft, 1922. 72 pp. (Ostwalds Klassiker der exakten Wissenschaften, Nr. 199). Translation: *Investigations on the Theory of the Brownian Movement*. (Translated by A. D. Cowper) London: Methuen, 1926, 122 pp. Reprint: New York, Dover Pubs., 1956.

Contains 5 papers by Einstein (1905, 1906, 1907, 1908) and extensive historical & technical notes by Furth.

DD4 [Einstein, A.] "Iz Prazhskikh Pisem A. Eǐnshteǐna - Iz Perepiski Eǐnshteǐna c M. Smolukhovskim." *Voprosy Istorii Estestvoznaniia i Tekhniki*, 18 (1965): 20-22.

From the correspondence of Einstein with M. Smoluchowski first published in Czechoslovakia in 1962 by A. Teske.

DD5 Furth, Reinhold. "Bericht ueber neuere Untersuchungen auf dem Gebiete der Brownsche Bewegung." *Jahrbuch der Radioaktivität und Elektronik*, 16 (1919): 319-61.

Review and bibliography of research on Brownian movement, 1913-19.

DD6 Herneck, Friedrich. "Zum Briefwechsel Albert Einsteins mit Ernst Mach." *Forschungen und Fortschritte*, 37 (1963): 239-43.

Einstein sends Mach his papers on Brownian movement.

DD7 Jost, Res. "Boltzmann und Planck: Die Krise des Atomismus un die Jahrhundertwende und ihre Überwindung durch Einstein." _Einstein Symposium, Berlin_ (item BE68), pp. 128-45.

DD8 Perrin, Jean. "Discontinuous structure of matter." _Nobel Lectures.Physics_ (item A87), vol. 2, pp. 138-64. December 11, 1926.

DD9 Teske, Armin. _The History of Physics and the Philosophy of Science: Selected Essays_. Wroclaw: Ossolineum, 1972. 210 pp.

Includes essays on Brownian movement.

* * * *

For additional references see HW, section D.hh.

DE. FLUID MECHANICS, HYDRODYNAMICS

DE1 Betz, Albert, et al. Hydro - and aerodynamics (FIAT Review of German Science, 1939-1946). Wiesbaden: Office of Military Government for Germany, Field Information Agencies Technical, British, French, U.S., 1948. 221 pp.

DE2 Böhme, Gernot. "Autonomisierung und Finalisierung." Starnberger Studien I: Die gesellschaftliche Orientierung des Wissenschaftlichen Fortschritte. Frankfurt: Suhrkamp Verlag, 1978, pp. 69-130.

DE3 Burgers, J. M. "Some memories of early work in fluid dynamics at the technical university of Delft." Annual Review of Fluid Mechanics, 7 (1975): 1-11.

* (Dryden, Hugh L.) Smith, Richard K. The Hugh L. Dryden papers, 1898-1965. Cited above as item BD10.

DE4 Greenspan, Harvey P. "This Week's Citation Classic: Greenspan, H. P. & L. N. Howard. On a time-dependent motion of a rotating fluid. J. Fluid Mech. 17: 385-404, 1963." Current Contents, Physical, Chemical & Earth Sciences, 21 no. 15 (April 13, 1981): 20.

The author recals the circumstances of writing this paper, which has been cited more than 135 times since 1963.

DE5 Kappler, Eugen, et al. Physics of Liquids and Gases (FIAT Review of German Science, 1939-1946) Wiesbaden: Office of Military Government for Germany, Field Information Agencies Technical, British, French, U. S., 1948. 348 pp. (Text in German)

DE6 Merkulova, N. M. Razvitie Gazovoi Dinamiki v SSSR. Moskva: Nauka, 1966. 151 pp.

Development of gas dynamics in USSR.

DE7 Oka, Syoten. "The physics of blood and blood vessels. Recent advances in haemorheology." Profiles of Japanese science and scientists. 1970. Edited by H. Yukawa. Tokyo: Kodansha, 1970, pp. 152-73.

With a biographical sketch of Syoten Oka, p. 151.

DE8 Truesdell, C. "The mechanical foundations of elasticity and fluid dynamics." Journal for Rational Mechanics and Analysis, 1 (1952): 125-300; 2 (1953): 593-616.

Critical-historical review of all major contributions to the subject in the past two centuries.

DE9 Truesdell, C. "The meaning of viscometry in fluid dynamics." _Annual Reviews of Fluid Mechanics_, 6 (1974): 111-46.

Survey of theory and experiment since about 1950.

DE10 Velarde, Manuel G. "Hydrodynamic instabilities (in isotropic fluids)." _Fluid Dynamics_. Edited by R. Balian and J. L. Peube. New York: Gordon & Breach, 1977, pp. 469-527.

Includes historical review and discussion of experiments of H. Benard (1900), M. J. Block (1956), etc.

* * * * *

For further references see HW, section D.e

DF. ACOUSTICS

DF1 Beyer, Robert. "A new wave of acoustics." Physics Today, 34, no. 11 (Nov. 1981): 145-57.

Survey of some developments since about 1950.

DF2 Biquard, P. "Les premiers pas dans les recherches sur les ultrasons." Journal de Physique, Vol. 33 (1972): C61-3

Discusses the first efforts by P. Langevin in the period 1914-1920 to generate and detect ultra-sound, and technical difficulties of the epoch.

DF3 Korpel, Adrianus. "Acousto-optics -- A Review of Fundamentals." Proceedings of the IEEE, 69,no. 1 (January 1981): 48-53.

Historical development of acousto-optics and its applications.

DF4 Lindsay, R. Bruce, Ed. Physical Acoustics. Stroudsburg, PA: Dowden, Hutchinson & Ross, 1974. xiii + 480 pp.

Includes reprints or translations of papers by A. G. Webster (1919), R. K. Cook and J. M. Young (1962), G. W. Stewart (1922), W. P. Mason (1927), R. B. Lindsay & F. E. White (1932), C. Eckart (1948), F. E. Borgnis (1953), P. J. Westervelt (1957), J. L. S. Bellin & R.T. Beyer (1962), W. G.Cady (1922), R.W. Wood & A. L. Loomis (1927), A. Einstein (1920), G. W. Pierce (1925), K. F. Herzfeld & F. O. Rice (1928), H. O. Kneser (1931), P. S. H. Henry (1932), V. O. Knudsen (1931, 1933), J. J. Markham, R. T. Beyer and R. B. Lindsay (1951), H. E. Boemmell and K. Dransfeld (1958), P. Debye and F. W. Sears (1932), L. Brillouin (1914), W. P. Mason and H. J. McSkimm (1947), W. P. Mason (1955), W. G. Proctor and W. H. Tanttilla (1955). Also, papers on liquid helium II by L. Tisza (1938), L. D. Landau (1941), K. R. Atkins (1959), I. Rudnick and K. A. Shapiro (1962), K. A. Shapiro and I. Rudnick (1965).

DF5 Nemchinova, N. M. "Iz Istorii Razvitii͡a Molekuliarnoĭ Akustiki." Istoriia i Metodologii͡a Estestvennykh Nauk, 2 (1963): 279-283.

Development of molecular acoustics at the turn of the century.

DF6 Nozdrev, V. F., A. A. Senkevich. "Razvitie Molekuli͡arnoĭ Akustiki v Sovetskom Soiuze za 50 Let." *Primenenie Ul'traakustiki k Issledovani͡iu Veshchestva. Moskovskii Oblastnoĭ Pedagogicheskiĭ Institut*, 23 (1967): 3-20.

Development of molecular acoustics in USSR from 1917 to 1967.

DF7 Shpakovskii, B. G. "Istoriia Komissii po Akustike AN SSSR (1935-1960)." *Trudy Komissii po Akustike, Akademiya Nauk SSSR*, 9 (1960): 6-20.

DF8 Slavin, I. I. "Kratkii ocherk rabot po issledovaniiu shumov i po bor'be s shumnostiiu v SSSR za period do 1957 g." *Akusticheskii Zhurnal*, 2(1958): 221-230.

Soviet research on and fight against noise.

DF9 "Sovetskaia Akustika Za 40 Let." *Akusticheskii Zhurnal* 4 (1957): 299-321.

Soviet Acoustics from 1917-1957. Editorial.

* * * *

For additional references see HW, section D.k

DG. SOLID STATE

DG1 Amsel, G. "This Week's Citation classic: Amsel, G, Nadal J P, d'Artemare E, David, D, Girard E & Moulin J. Microanalysis by the direct observation of nuclear reactions using a 2 MeV Van de Graaff. Nucl. Instrum. Methods 92: 481-98, 1971." Current Contents, Physical, Chemical & Earth Sciences, 20, no. 33 (August 18, 1980): 12.

The author recalls the circumstances of writing this paper, which has been cited more than 165 times since 1971.

DG2 Andresen, A. F. "20 Years of Polarized Neutron Research." Nukleonika, 24 (1979): 691-8.

Looks back on 1957 experiment which started the technique of polarized neutron diffraction to study magnetic phenomena. Describes previously unpublished details on the experimental approach to polarized neutrons using crystal diffraction.

DG3 Auth, J. "Quantentheorie und Halbleitertechnik." 75 Jahre Quantentheorie. Edited by W. Brauer W. et al. Abhandlungen der Akademie der Wissenschaftender DDR, Abt.Math.Naturwiss. Technik, 1977, Nr.7N, pp. 115-132.

DG4 Balarin, M., ed. Festkoerperphysik-Entwicklungstendenzen und Anwendungsmoeglichkeiten. Berlin: Akademie-Verlag, 1976. 252 pp. (Reihe WTB Texte und Studien, Bd. 168)

Contains item DG21.

DG5 Bardeen, John. "Semiconductor research leading to the point contact transistor." Nobel Lectures...Physics (item A87), vol. 3, pp. 318-41. December 11, 1956.

DG6 Bragg, W. L. "Forty Years of Crystal Physics." Background to Modern Science (item A85), pp. 77-89.

DG7 Brattain, Walter H. "Surface properties of semiconductors." Nobel Lectures...Physics (item A87), vol. 3, 377-84. December 11, 1956.

DG8 Brauer, W. "Die Physik der Metalle und ihre Beeinflussung durch die Quantentheorie." 75 Jahre Quantentheorie. Edited by W. Brauer et al. Abhandlung der Akademie Wissenschafter.DDR, Abt. Math Naturwiss Technik, (1977), Nr.7N, pp. 101-113.

DG9 Braun, Ernest and Stuart MacDonald. Revolution in Miniature: The History and Impact of Semiconductor Electronics. New York: Cambridge University Press, 1978. 232 pp.

* Brush, Stephen G. "History of the Lenz-Ising Model." Cited above as item DB6.

DG10 Busch, Georg. "Festkoerperphysik-wohin?" _Physikalische Blaetter_, 37 (1981): 333-6.

DG11 Carruthers, Peter. "This week's citation classic: Carruthers, P. Theory of thermal conductivity of solids at low temperatures. _Rev. Mod. Phys_. 33: 92-138, 1961." _Current Contents, Physical, Chemical & Earth Sciences_, 20, no. 45 (10 Nov. 1980): 14.

The author recalls the circumstances under which he wrote this paper, which has been cited over 235 times since 1961.

DG12 Casimir, H. B. G. "Development of Solid-State Physics." _History of Twentieth Century Physics_ (item A140), pp. 158-69.

DG13 Cervinka, L. "Zánik trojhvězdí anglických krystalografů." (I, II, III) _Československý Časopis pro fyziku_, sekce A, 22 (1972): 306-8, 424-26, 531-32.

The doom of three stars of English crystallography: K. Lonsdale, L. Bragg, J. D. Bernal.

* Chaudhari, Praveen. "This Week's Citation Classic: Chaudhari, P. et al. Amorphous metallic films for bubble domain applications..." Cited below as item E4.

DG14 Chu, Wei-Kan. "This Week's Citation Classic: Chu W K, Mayer J W, Nicolet M A, Buck T M, Amsel G & Elsen F. Principles and applications of ion beam techniques for the analysis of solids and thin films. _Thin Solid Film_ 17: 1-41, 1973." _Current Contents, Physical, Chemical & Earth Sciences_, 20, no. 35 (September 1, 1980), 12.

The author recalls the circumstances of writing this paper, which has been cited more than 195 times since 1973.

DG15 Cowley, R. A. "This Week's Citation Classic: Cowley R A. The lattice dynamics of an anharmonic crystal. _Adva. Phys_. 12: 421-80, 1963." _Current Contents, Physical, Chemical & Earth Sciences_, 21, no. 42 (October 19, 1981): 20.

The author recalls the circumstances of writing this paper, which has been cited more than 350 times since 1963.

DG16 Damask, A. C. "This Week's Citation Classic: Damask A C & Dienes G J. Point defects in metals. New York: Gordon and Breach, 1963. 314 pp." Current Contents, Physical, Chemical & Earth Sciences, 21, no. 39 (September 28, 1981): 20.

The author recalls the circumstances of writing this paper, which has been cited more than 630 times since 1963.

DG17 Debye, P. "The Solid State around 1910."" Ferroelectricity. Edited by Eduard F.Weller. Amsterdam: Elsevier Publishing Co., 1967, pp. 297-303.

DG18 Dubchak, V. A. and V. M. Konovalov. "Osnovnye Etapy Razvitiia Zonnoĭ Teorii Tverdykh Tel." Istoriia i Metodologiia Estestvenykh Nauk, 4 (1966): 220-230.

On the theory of "zones" in solid state physics developed in the years 1924-28.

DG19 El'iashevich, M. A. and T. S. Prot'ko. "Eĭnshteĭn i Kvantovaia Fizika Tverdogo Tela." Einshtein i Razvitie Fiziki. Edited by G. M. Edlis. Moskva, 1981.

Einstein and quantum solid state physics.

DG20 Esaki, Leo. "Long Journey into Tunneling." Proceedings of the IEEE, vol. 62, 6 (June 1974): 825-31.

Discussion of historical background of quantum theory of tunneling.

* Fradlin, B. N. and V. M. Sliusarenko. "O nekotory issledovaniiakh otechestvennykh uchenykh po dinamiko tverdogo tela." Cited above as item D1.

DG21 Frankfurt, U. J. and A. M. Frenk. "Der Festkorper." Festkoerperphysik, Entwicklungstendenzen und Anwendungsmoeglichkeiten (item DG4), pp. 68-82.

DG22 Frenkel, V. Ia. "Iz istorii prilozheniia kvantovomekhanicheskikh predstavlenii k teorii svobodnykh elektronov." Voprosy Istorii Estestvoznaniia i Tekhniki, no. 3-4 (56-57) (1977): 57-62.

On the history of the application of quantum-mechanical concepts to the theory of free electrons, 1926-30; discusses the work of J. Frenkel, L. D. Landau, J. G. Dorfman, M. P. Bronstein and I. E. Tamm.

DG23 Frenkel, V. Ya. Semiconductors, USSR Academy of Sciences, The Lenin Order A.F. Ioffe Physicotechnical Institute, Leningrad: Nauka Publishers, Leningrad Branch, 1979. 135 pp. Chief Editor, Acad. V.M. Tuchkevich; compiled by V. Ya. Frenkel.

This booklet is printed in connection with the 60th anniversary of the A. F. Ioffe Physicotechnical Institute. It gives a historical survey of semiconductor research at the Institute during 1930-1968 and a detailed acount of data obtained over the last ten years. This account covers a wide range of research problems, including semiconductor devices and materials for them, the physics of heterojunctions and excitons, semiconductors, nonequilibrium processes, and some other questions.

DG24 Friedel, J. "This week's citation classic: Friedel J. Metallic alloys. Nuovo Cimento 7: 287-311, 1958." Current Contents, Physical, Chemical & Earth Sciences, 20, no. 49 (8 Dec. 1980): 10.

The author recalls his work on the electronic structure of alloys which resulted in this paper, cited over 520 times since 1961.

DG25 Gatos, H. C. "Structure Property Relationships: Key to Solid-State Science and Technology." Journal of the Electrochemical Society, 122 (1975): 287c-300.

Views of the relationship between solid-state and technology in a historical framework. Discussion of evolution of a few research themes.

DG26 Geballe, Theodore H. "The golden age of solid-state physics." Physics Today, 34, no. 11 (Nov. 1981): 132-43.

Historical survey, concentrating on developments since 1955.

DG27 Goetzeler, Herbert. "Zur Geschichte der Halbleiter-Bausteine der Elektronik." Technikgeschichte, 39 (1972): 31-50.

DG28 Goodman, Alvin M. "This Week's Citation Classic: Goodman A M. Metal-semiconductor barrier height measurement by the differential capacitance method - one carrier system. J. Appl. Phys. 34: 329-38, 1963." Current Contents, Physical, Chemical & Earth Sciences, 21, no. 43 (October 26, 1981), 20.

The author recalls the circumstances of writing this paper, which has been cited more than 170 times since 1963.

DG29 Grau, Monika. _Die Anfaenge der Festkoerperphysik in Stuttgart_. Stuttgart: Historisches Institut, Zulassungsarbeit, 1979.

DG30 Guentheroth , Horst. "Die Wiege der modernen Halbleiterphysik." _Bild der Wissenschaft_, 5 (1979): 170-172.

On the history of Hall's effect.

DG31 Guillaume, Charles-É. "Invar and elinvar." _Nobel Lectures...Physics_ (item A87), vol. I, pp. 444-73. December 11, 1920.

DG32 Heywang, W. and E. Spenke. "Twenty Five Years of Semiconductor-Grade Silicon." _Physica Status Solidi_ (a) 64 (1981): 11-44.

DG33 Heywang, Walter. "Festkoerperphysik -- Quo vadis." _Physikalische Blaetter_, 37 (1981): 152-53.

DG34 Hillert, Mats. "This Week's Citation Classic: Hillert M. A solid-solution model for inhomogeneous systems. _Acta Metallurgica_ 9: 525-35, 1961." _Current Contents, Engineering, Technology & Applied Sciences_, 12, no. 33 (August 17, 1981): 18.

The author recalls the circumstances of writing this paper, which has been cited more than 155 times since 1961.

DG35 Hirth, J. P. "This Week's Citation Classic: Hirth J P & Lothe J. _Theory of dislocations_. New York: McGraw-Hill, 1968. 780pp." _Current Contents, Physical, Chemical& Earth Sciences_, 21, no. 25 (June 22, 1981): 18.

The author recalls the circumstances of writing this book, which has been cited more than 1,040 times since 1968.

DG36 Hoddeson, Lillian and Nobuo Kawamiya. "Japanese-American Symposium on the History of Modern solid-state physics, Nagoya, Japan, 2-4 December, 1979." _Isis_, 71 (1980); 628-29.

DG37 Hoddeson, Lillian H. and G. Baym. "The development of the quantum mechanical electron theory of metals: 1900-28."" _Proceedings of the Royal Society of London_, A 371 (1980): 8-23.

On the work of W. Pauli, A. Sommerfeld, F. Bloch etc.

DG38 Hoddeson, Lillian. "The discovery of the point-contact transistor." _Historical Studies in the Physical Sciences_, 12 (1981): 41-76.

DG39 Hoddeson, Lillian Hartman. "The entry of the quantum theory of solids into the Bell Telephone Laboratories, 1925-1940: A case-study of the industrial application of fundamental science." _Minerva_, 18 (1980, pub. 1982): 422-47.

DG40 Hubbard, J. "This Week's Citation Classic: Hubbard J. Electron correlations in narrow energy bands. _Proc. Roy. Soc. London. Ser._ A 276 (1963): 238-57." _Current Contents, Physical, Chemical & Earth Sciences_, 20, no. 22 (June 22, 1980): 16.

The author recalls the circumstances of writing this paper, which has been cited more than 855 times since 1963.

DG41 Hulin, Michel. "La Physique du Solide. Son role dans la genese des idees quantiques." _C.U.I.D.E._ (Universite Pierre et Marie Curie) 6, no. 19(Avril 1981): 1-41.

* Hulin, M. "'En attendant Debye..'" Cited below as item HB34.

DG42 Ibers, James A. "This Week's Citation Classic: Ibers J A & Hamilton W C. Dispersion corrections and crystal structure refinements. _Acta Crystallogr._ 17: 781-2, 1964." _Current Contents, Physical, Chemical, & Earth Sciences_, 20, no. 39 (September 29, 1980): 22.

The author recalls the circumstances of writing this paper, which has been cited more than 345 times since 1964.

DG43 _IEEE Transactions on electron devices_. Special issue. "Historical notes on important tubes and semiconductor devices." Vol. ED-23, Nr. 7, July 1976

DG44 Joos, Georg, et al. _Physics of solids_. (FIAT Review of German science, 1949-1946) Wiesbaden: Office of Military Government for Germany, Field Information Agencies Technical, British, French, U.S., 1947.

DG45 Kahng, D. "A Historical Perspective on the Development of MOS Transistors and Related Devices." _IEEE Transactions on Electron Devices_, Vol. ED-23, N.7 (July 1976): 655-7

DG46 Kikoin, I. K. and Iu. N. Smirnov. "Puti Razvitiia Fiziki Tverdogo Tela." _Priroda_, 2 (1968): 2-12.

DG47 Koptsik, V. A. and I. S. Rez. "Raboty P'era Kiuri v oblasti kristallofiziki." Uspekhi Fizicheskikh Nauk, 134 (1981): 149-52.

Pierre Curie's work in crystal physics.

DG48 Langenberg, D. N. "Resource Letter OEPM-1 on the Ordinary Electronic Properties of metals." American Journal of Physics, 36 (1968): 777-78.

Includes some historical references.

DG49 Morandi, G., F. Napoli and C. F. Ratto. "The Friedel-Anderson model and the dynamics of contemporary research." Fundamenta Scientiae, 1 (1980): 233-59.

On the work of J. Friedel (1952), P. W. Anderson (1961) and others; model for formation of localized magnetic moments in dilute metallic alloys.

DG50 Mott, Sir Nevil. "Sixth Hume-Rothery Memorial Lecture. Electrons in crystalline and non-crystaline metals - from Hume-Rothery to the present day." Metal Science, 14 (12) (1980): 557-561.

DG51 Mott, Nevill. "Theory and experiment since Schroedinger's equation." Physics Bulletin, 25 (1974): 448-51.

DG52 Mott, Sir Nevill. "Atoms in Contact I: components of the solid state. The basic ideas of solid state physics were formulated in the hectic decades of the 1920s and 1930s." New Scientist, 69 (1976): 663-666.

DG53 Mott, Sir Nevill. "Atoms in contact II: Age of development." New Scientist, 69 (1976): 997, 172-74.

Applications to semi conductors. In Britain theorists concerned themselves with metals.

DG54 Mott, N. F., et al. "The Beginnings of Solid State Physics. A Symposium held 30 April - 2 May 1979." Proceedings of the Royal Society of London, A 371 (1980): 1-177.

Includes item DG38.

DG55 Mott, Nevill. "This week's citation classic: Mott N F. Electrons in disordered structures. Advan. Phys. 16: 49-144, 1967." Current Contents, Physical, Chemical & Earth Sciences, 20, no. 51 (22 Dec. 1980): 18.

The author recalls the writing of this paper and his other contributions to the subject. The paper has been cited over 490 times since 1967.

DG56 Nagaoka, Y. "Theory of electrons in metals." Solid State Physics (Japan), 2, no. 2 (Nov 1976): 615-624.

Brief history (in Japanese) of the theory of electrons with discussion of why a simple free electron model gives results in good agreement with experiment.

DG57 Pake, George E. "This Week's Citation Classic: Pake, G. E. Nuclear resonance absorption in hydrated crystals: fine structure of the proton line. J. Chem. Phys. 16: 327-36, 1948." Current Contents, Physical, Chemical & Earth Sciences, 21, no. 49 (December 7, 1981): 16.

DG58 Peierls, R. E. "Quantum theory of solids." Theoretical Physics in the Twentieth Century (item BP4), pp. 140-60.

See also the article by H. B. G. Casimir in the same book.

DG59 Pick, H. "Fifty years of Colour Centre Physics." Journal de Physique, Colloque C6, Supplement au No.7, Tome 41, (Juillet 1980): C6-1--C6-6.

DG60 Richardson, Owen W. "Thermionic phenomena and the laws which govern them." Nobel Lectures..Physics (item A87), vol. 2, pp. 224-36. December 12, 1929.

DG61 Rupp, H. and G. Walter, Eds. Generation and Power Amplification in modern relay systems using all solid state devices. Symposium on Microwave Solid State Devices and Applications, Johannesburg, South Africa, 11 May 1977. Transactions of the South African Institute of Electrical Engineers, 71, Pt. 3 (March 1980): 61-6.

Traces the development of microwave semiconductors that brought about the disappearance of tubes from radio relay systems.

DG62 Schomaker, Verner. "This Week's Citation Classic: Schomaker V & Trueblood K N. On the rigid-body motion of molecules in crystals. Acta Crystallogr. B24: 63-76, 1968." Current Contents, Physical, Chemical & Earth Sciences, 20, no. 31 (August 4, 1980): 14.

The author recalls the circumstances of writing this paper, which has been cited more than 735 times since 1968.

DG63 Seitz, Frederick. "Festkoerperphysik-Gestern, Heute, Morgen." Physik in unserer Zeit, 6, Nr. 1 (1979): 15-24.

DG64 Shockley, William. "Transistor technology evokes new physics." Nobel Lectures...Physics (item A87), vol. 3, pp. 344-74. December 11, 1956.

DG65 Slater, John C. "Quantum Physics in America between the Wars." "Energy Bands in Solids." Physics Today, 21, no. 1 (Jan. 1968): 43-51; no. 4 (April 1968): 61-71.

Two-part historical account of the Wigner-Seitz method for interpreting metallic structure, study of Brillouin zones and calculation of energy bands in solids, and postwar research facilitated by commercial interest in transistors and availability of digital computers.

DG66 Sodomka, L. "Z historie fyziky pevnych latek." Matematika a fyzika ve skole, 6 (1975-76): 288-91.

"From the history of solid-state physics."

DG67 Stein, Dale F. "This week's citation classic: Stein D F & Low J R. Mobility of edge dislocations in silicon-iron crystals. J. Appl. Phys. 31: 262-9, 1960." Current Contents, Physical, Chemical & Earth Sciences, 20, no. 32 (11 Aug. 1980): 18.

The first author recalls the original research and preparation of this paper, which has been cited over 290 times since 1961.

DG68 Suleimanian, G. A. "K istorii pervonachal'nykh primeneniĭ kvantovykh printsipov v teorii elektroprovodnosti metallov.." Voprosy istorii Estestvoznaniia i Tekhniki, 3 (1981): 95-99.

On the history of the application of quantum principles in metal conduction theory.

DG69 Suleimanian, G. A. "Eĭnshteĭn i Teoriia Elektroprovodnosti Metallov." Eĭnshteĭn i Razvitie Fiziki. G. M. Idlis, Ed. Moskva, 1981.

Einstein and the theory of electron conduction of metals.

DG70 Taylor, Geoffrey. "Note on the early stages of dislocation theory." The Sorby Centennial Symposium on the History of Metallurgy. Edited by C. St. Smith. New York: Gordon and Breach Science Publishers, 1965, pp. 355-358.

DG71 Uyeda, R. "Electron Diffraction and Microscopy in Japan past and present." Journal of Applied Crystallography, 7, Pt. 1 (February, 1974): 1-18.

Lecture begins with the experiment of Kikuchi and relates work under Miyake and Uyeda.

DG72 Weiss, H. "Steuerung von Elektronenstroemen im Festkoerper." Physikalische Blaetter, 31, no. 4 (April 1975): 156-165.

Lecture delivered on the occasion of 60th birthday of Heinrich Welker with an account of his contribution (late '20s and early '30s) to the elucidation of the decisive relationship governing the control by electric or magnetic fields of electron flow in solid and the bipolar transistor.

DG73 Welker, Henrich. "Discovery and Development of III-V-Compounds." IEEE Transactions on Electron Devices, Vol. ED-23, Nr. 7 (1976): 664-674.

DG74 Welker, Heinrich J. "From solid state research to semiconductor electronics." Annual Review of Materials Science, 9 (1979): 1-21.

History of semiconductor research from its origins.

* * * * *

For additional references see HW, section O.

DH. LOW TEMPERATURE PHYSICS

DH1 Anderson, P. W. "How Josephson discovered his effect." Physics Today, 23, no. 11 (Nov. 1970): 23-29.

Discovery of quantum tunnelling in superconductors by Brian Josephson (1962) and some of its applications.

DH2 Balabekian, O. I. "K Istorii Razvitiia Fiziki Nizkikh Temperatur v SSSR." Uchenye Zapiski Orenb. Politekhnicheskii Institut (1967): 140-155.

On the development of low temperature physics in the USSR.

DH3 Bardeen, John. "Electron-phonon interactions and superconductivity." Physics Today, vol. 26, no. 7 (July 1973): 41-47.

Nobel Lecture.

DH4 Brandt, N. B. "Fizika nizkikh temperatur." Razvitie Fizike v Rossii (item A93), pp. 117-55.

DH5 Casimir, H. B. G. "Superconduction in the 1930s." (In Dutch) Nederlands Tijdschrift voor Natuurkunde., A46, no. 4 (Dec. 1980): 137-40.

Describes effect of quantum mechanics (1924-27) on research work during early 1930s culminating in the "two-fluid theory". Reference to C. J. Gorter's work on thermodynamics of superconductors.

DH6 Casimir, H. B. G. "Superconductivity and superfluidity." The Physicist's Conception of Nature (item A80), pp. 481-98.

DH7 Cooper, Leon. "Microscopic quantum interference in the theory of superconductivity." Science, 181 (1973): 908-16.

Nobel Lecture.

DH8 Czerwonko, J. "Superfluid phases of ^3He." Postery Fizyki., 29, no. 4 (1978): 387-99.

DH9 Dobrov, G. M. and I. V. Dziekovskaya. "Methods and results of studying the flow of information in the field of thin-film superconductivity." Scientometrics, 4 (1982): 27-44.

Statistical analysis of the literature from 1949-77. The problem of professional mobility is also investigated.

DH10 Galasiewicz, Zygmunt M. *Helium 4*. New York: Pergamon Press, 1971. vii + 338 pp.

Includes papers by H. Kamerlingh Onnes & J. D. A. Boks (1924), W. H. Keesom & M. Wolkfe (1927), W. H. Keesom & K. Clusius (1932), J. F. Allen, R. Peierls & M. Z. Uddin (1937), J. F. Allen & H. Jones (1938), P. Kapitza (1938, 1941), W. H. Keesom & G. E. MacWood (1938), J. G. Daunt & K. Mendelssohn (1938), E. Andronikashvili (1946), V. Peshkov (1946), L. D. Landau (1941, 1947), E. Lifshitz (1944), N. Bogolubov (1947), R. P. Feynman (1955), I. M. Khalatnikov (1952), L. P. Pitaevskii (1956).

DH11 Giauque, William F. "Some consequences of low temperature research in chemical thermodynamics." *Nobel Lectures. Chemistry* (item A86), vol. 3, pp.227-50. December 12, 1949.

DH12 Gorter, C. J. "Superconductivity until 1940 in Leiden and as seen from there." *Reviews of Modern Physics*, 36 (1964): 3-7.

DH13 Hallock, Robert B. "Resource Letter SH-1: Superfluid helium." *American Journal of Physics*, 50 (1982): 202-12.

Annotated bibliography of works suitable for teachers and students.

DH14 Hudson, R. P. " A century of cryogenics." *Journal of the Washington Academy of Sciences*, 67 (1977): 119-30.

DH15 Hulm, John K., Eugene Kunzler and Bernd T. Matthias. "The road to superconducting Materials." *Physics Today*, 34, no. 1 (January 1981): 34-43.

DH16 Iavelov, B. E. "O Statie A. Eĭnshteĭna 'Teoreticheskie Zamechanii͡a K Sverkhprovodimosti Metallov'." *Voprosy Istorii Estestvoznanii͡a i Tekhniki*, 67-68 (1969): 46-53.

On Einstein's paper "Theoretical observations on metal supraconductivity."

DH17 Iavelov, B. E. "Eĭnshteĭn i problema sverkhprovodimosti." *Eĭnshteinovskiĭ Sbornik 1977*, pp. 158-86.

Einstein and the problem of superconductivity.

DH18 Kamerlingh Onnes, Heike. "Investigations into the properties of substances at low temperatures, which have led, amongst other things, to the preparation of liquid helium." *Nobel Lectures...Physics* (item A87), vol. I, pp. 306-36. December 11, 1913.

* Lindsay, R. Bruce, Ed. Physical Acoustics. Cited above as item DF54.

* Maglich, Bogdan, ed. Adventures in Experimental Physics. Cited above as item A74.

Includes accounts of the discovery of quantized circulation in superfluid, Josephson effect, electron tunneling into superconductors.

DH19 Mendelssohn, K. "Prewar work on Superconductivity as seen from Oxford." Reviews of Modern Physics, 36 (1964): 7-12.

DH20 Nadel, E. "Citation and co-citation indicators of a phased impact of the BCS theory in the physics of superconductivity." Scientometrics, 3 (1981): 203-21.

DH21 Nernst, Walther Hermann. "Studies in Chemical Thermodynamics." Nobel Lectures...Chemistry (item A86), Vol. I, pp. 353-62. December 12, 1921.

DH22 Pines, David. "Elementary excitations in quantum liquids." Physics Today, 34, no. 11 (Nov. 1981): 106-31.

Survey of developments since 1931.

DH23 Richardson, Robert C. "Low temperature science -- what remains for the physicist?" Physics Today, 34, no. 8 (Aug. 1981): 46-50.

On B. Pippard's 1961 lecture asserting that, apart from high energy physics, there were no more really fundamental questions remaining in 'pure' physics. With a comment by Pippard.

DH24 Pulvermueller, Siegfried. Geschichte der Supraleitung bis 1940. Stuttgart: Historisches Institut, Zulassungsarbeit, 1972.

DH25 Schmitt, R. W. "The discovery of electron tunneling into superconductors." Physics Today, 14, no. 12 (Dec. 1961): 38-41.

On the discovery by Ivar Giaever in 1961.

DH26 Schrieffer, J. Robert. "Macroscopic quantum phenomena from pairing in superconductors." Physics Today, 26, no. 7 (July 1973): 23-28.

Nobel Lecture.

* Trigg, George L. *Landmark Experiments in Twentieth Century Physics*. Cited below as item JD20.

Chapters on Superconductivity and liquid helium.

* * * * *

For additional references see HW, section O.h

DI. HIGH TEMPERATURE/PLASMA PHYSICS

DI1 Bostick, Winston U. "The pinch effect revisited." International Journal of Fusion Energy, 1 (1977): 1-53.

Historical review of controlled thermonuclear fusion.

DI2 Kabakova, A. I. "Nachal'nyi Period razvitiia teorii nizotemperaturnic plazmy." Voprosy Istorii Estest-voxnaniia i Tekhniki, 54, no. 1 (1976): 52-56.

"The initial stage of the formation of the theory of low-temperature plasma and the causes which moved this field of physics to the frontiers of science are ... considered. It is shown that the theory of low-temperature plasma was developed among two directions: in the works by V. Shostka and [I.] Langmuir." (Author's summary)

DI3 Kapitza, P. L. "Plasma and the controlled thermonuclear reaction." Science, 205 (1979).

Noble lecture.

* Tamm, Igor E. "General characteristics of radiations emitted by systems moving with super-light velocities with some applications to plasma physics." Cited below as item G45.

DI4 Tonks, Lewi. "The birth of 'plasma'." American Journal of Physics, 35 (1967): 857-58.

DI5 Trivelpiece, Alvin W. "This Week's Citation Classic: Trivelpiece A W & Gould R W. Space charge waves in cylindrical plasma columns. J. Appl. Phys. 30: 1784-93, 1959." Current Contents, Physical, Chemical & Earth Sciences, 21, no. 29 (July 20, 1981): 18.

The author recalls the circumstances of writing this paper, which has been cited more than 270 times since 1961.

DI6 Willson, Denis. A European Experiment. London: Adam Higler Ltd., 1981. 178 pp.

An account of JET (Joint European Torus) and the politics of international scientific collaboration and their adverse effect on the JET fusion energy experiment within the European community.

E. ELECTRICITY AND MAGNETISM

E1 Brown, W. Fuller. "Domains, Micromagnetics and Beyond: Reminiscences and Assessment." Journal of Applied Physics, 49 (3) (1978): 1937-42.

E2 Brush, Stephen G. "Francis Bitter and 'Landau diamagnetism'." Journal of Statistical Physics, 2 (1970): 195-97.

* Bruzzaniti, Giuseppe. "'Real history' as 'Dictionary' Reconstruction. A Historiographic Hypothesis for Pierre Curie's Scientific Undertaking." Cited above as item BC30.

E3 Butler, J. W. "On the Trouton-Noble experiment." American Journal of Physics, 36 (1968): 936-41.

E4 Chaudhari, Praveen. "This Week's Citation Classic: Chaudhari P, Cuomo J J & Gambino R J. Amorphous metallic films for bubble domain applications. IBM J. Res. Dev. 17: 66-8, 1973." Current Contents, Engineering, Technology, & Applied Sciences, 12, no. 45 (November 9, 1981): 16.

The author recalls the circumstances of writing this paper, which has been cited more than 165 times since 1973.

E5 Frenkel, Victor Ya. "K istorii effekta Eĭnshteĭna - De Gaaza." Uspekhi Fizicheskikh Nauk, 128 (1979): 545-557.

On the history of the Einstein-De Haas effect.

E6 Galison, Peter L. "Theoretical dispositions in Experimental Physics: Einstein and the Gyromagnetic experiments, 1915-1925." Historical Studies in the Physical Sciences, 12 (1982): 285-323.

E7 Hirosige, Tetu. "Electrodynamics before the theory of relativity, 1890-1905." Japanese Studies in History of Science, 5 (1966): 1-49.

E8 Ioffe, A. F. "I. V. Kurchatov-Issledovatel' Dielektrikov." Uspekhi Fizicheskikh Nauk, 73 (1961): 611-14.

E9 Kirenskii, L. V., R. V. Telesnin, and Ia. S. Shur. "Fizika Magnetizma." Razvitie Fiziki v Rossii (item A93), pp. 76-99.

E10 Lenard, Philipp. _Great Men of Science. A History of Scientific Progress_. (Translated from the Second German edition, 1932, by H.S. Hatfield) London: Bell, 1958. Reprint, Freeport, N.Y.: Books for Libraries Press, 1970. xi + 389 pp.

Includes a lengthy discussion of F. Hasenoehrl and his work on the mass of electromagnetic energy, with several references to Lenard's own theories of the ether, avoiding any mention of Einstein.

E11 Mesnage, P. "One hundred years of piezoelectricity Studies." _Bulletin Annuel de la Societe Suisse Chronometrie et du Laboratoire Suisse de Recherches Horlogeres_, 9 (1980): 165-168. In French.

History of piezoelectricity from the Curie brothers.

E12 Miller, Arthur I. "Unipolar induction: A case study of the interaction between science and technology." _Annals of Science_, 38 (1981): 155-89.

E13 Moon, Parry and Domina Eberle Spencer. "Electromagnetism without magnetism: An Historical Sketch." _American Journal of Physics_, 22 (1954): 120-24.

On Attempts to formulate electrodynamics as an extension of Coulomb's equation, eliminating the magnetic field, by Coulomb, Ampere, Gauss, W. Ritz (1908), F. W. Warburton (1946), and the authors (1953).

E14 Petruccioli, Sandro. "La dialettica teorica della fisica prerelativistica e il 'programma elettromagnetico' di Max Abraham." _Seminario di Storia della Fisica_ (item A110), pp. 31-51.

E15 Pourprix, Bernard. "L'evolution des idees sur la polarisation dielectrique." CUIDE 6, no. 18 (Jan. 1981): 47-76 (published by Universite Pierre et Marie Curie).

* Richardson, Owen, W. "Thermionic phenomena and the laws which govern them." Cited above as item DG60.

E16 Rytov, S. M. "Razvitie teorii nelineinykh kolebanii v SSSR." _Radiotekhnika i Elektronika_, 11 (1957): 1435-50.

Covers developments in theory of nonlinear vibrations in preceding 40 years in USSR with extensive bibliography.

E17 Stockmann, F. "Photoconductivity - A Centennial." *Physica Status Solidi A*, 15 (1973): 381-90.

Discussion of developments from 1879-1945 and 1946-1972 and applications.

E18 Suesskind, Charles. "American Contributions to Electronics: Coming of age and some more." *Proceedings of the IEEE*, 64 (1976): 1300-5.

Rise of a "science intensive" industry, electronics, illustrates the transition of US froma scientifically underveloped country to a world leader in science.

E19 Tarsitani, Carlo. "Lorentz e l'autonomia teorica del concetto di campo elettromagnetico." *Seminario di Storia della Fisica* (item A110), pp. 9-30.

E20 Valasek, J. "The Early History of Ferroelectricity." *Ferroelectrics*, 2, (1971): 239-44.

Traces the history of ferroelectricity up to the discovery of the ferroelectric properties of barium titanate in 1943.

E21 Vleck, J. H. Van. "Quantum mechanics - The key to understanding magnetism." *Reviews of Modern Physics*, 50 (1978): 181-89.

Nobel Lecture, 1977.

E22 Vleck, J. H. Van. "The widening world of magnetism." *Physics Bulletin*, 19 (1968): 167-75.

Survey of developments in the past 50 years.

* * * *

For additional references see HW, section E.

F. RELATIVITY - General and Miscellaneous

F1 Belloni, Lanfranco. "Proposta di letture di storia della fisica." _Museoscienza_, 3 (Luglio-Sett. 1981): 11-14.

Includes a 1923 note by Fermi on mass in relativity theory.

F2 Birkhoff, George David. _The origin, nature and influence of relativity._ New York: Macmillan, 1925. ix + 185 pp.

F3 Biro, Gabor. "Gyözö Zemplén a přijetí teorie relativity v Madarsku." _Práce z dejin prírodnich věd_, 13 (1980): 47-53.

Gyözö Zemplan and the acceptance of the theory of relativity in Hungary.

F4 Cajori, F. "Are the heavens full or are they void? A History of hypotheses." _Scientific Monthly_, 23 (1926): 346-55.

F5 Caldirola, Piero. "Progressi nella teoria dell 'elettrone." _Annuario della EST 1979_ (Enciclopedia della Scienza e della Tecnica) Milano: Mondadori, 1979, pp. 66-72.

F6 Einstein, Albert. "How I created the Theory of Relativity." _Physics Today_, 35, no. 6 (August 1982): 45-47.

Translation by Y. A. Ono of notes by J. Ishiwara on a lecture given by Einstein in Kyoto, 14 December 1922.

F7 Einstein, Albert. "Fundamental ideas and problems of the theory of relativity." _Nobel Lectures...Physics_ (item A87), Vol. 1, pp. 482-90. Delivered to the Nordic Assembly of Naturalists at Gothenburg, July 11, 1923.

The lecture was not delivered on the occasion of the Nobel Prize Award, and did not, therefore, concern the discovery of the photoelectric effect.

F8 Gruenbaum, Adolf. _Philosophical Problems of Space and Time_. Second Edition. (Boston Studies in the Philosophy of Science, 12) Dordrecht: Reidel, 1973. xxiii + 884 pp.

Gruenbaum's discussion of the history of special relativity is criticized by Arthur Miller in his review in _Isis_, 66 (1975): 590-94. See reply by Gruenbaum in _Isis_, 68 (1977): 447-48 and Miller's rejoinder, _ibid._, 449-50.

* Hermann, Armin and Ulrich Benz. "Quanten- und Relativitaetstheorie im Spiegel der Naturforscherversammlungen 1902-1920." Cited above as item H16.

F9 Holton, Gerald. "The metaphor of space-time events in science." Eranos-Jahrbuch 34 (1965): 33-78.

F10 Illy, Joszef. "Lenin, elektromagnetický obraz světa a teorie relativity." Práce z dějin přírodních věd, 13 (1980): 31-38.

Lenin, the electromagnetic world view and the theory of relativity.

* Kagan, Martin H. and Etic Mendoza. "Down with the History of Relativity!" Cited below as item M36.

F11 Kuznetsov, B. G. Printsip Otnositel'nosti v Antichnoĭ, Klassicheskoĭ i Kvantovoĭ Fizike. Moskva: Izd-vo AN SSR, 1959. 232 pp.

Priniciple of relativity in ancient, classical and quantum physics.

F12 Kuznetsov, B. G. Veseli za teoriyata na otnositelnostta. Sofia: Tekhnika, 1962. 201 pp. Translations: O teorii relativity, Praha: Statni nakl. technické literatury, 1962. 237 pp.; Teoria Wzgledności, Warzsawa: Panstw. wyd-wo noukowe, 1962. 254 pp.; Gesprekken over de relativiteits theorie, Antwerpen: Het Spectrum, 1967. 236 pp.; Essais sur la relativite, Moscow: "Mir" 1979. 266 pp.; Saggio sulla relativita, Roma: Tiuniti/Moscow: Mir, 1974. 286 pp.

F13 Laemmel, Rudolf. Ideile lui Einstein si Teoria relativitătii pe întelesul tuturor. Translated by I. Radŭlescu. Second Edition. Bucuresti: Ed. Minescu. 104 pp.

F14 Lanczos, Cornelius. Space through the ages: The evolution of geometrical ideas from Pythagoras to Hilbert and Einstein. New York: Academic Press, 1970. x + 320 pp.

F15 Markov, Moiseĭ Aleksandrovich. "S. I. Vavilov i teorii͡a otnositel'nosti." Voprosy Istorii Estestvoznaniia i Tekhniki, 3 (1981): 37-39.

S. I. Vavilov and relativity theory.

F16 Pantaleo, Mario. Ed. Cinquant'anni di relativita, 1905-1955. Second edition. Firenze: Edizioni Giuntini & Sansoni Editore, 1955. 634 pp.

Includes essays by G. Polvani, P. Straneo, B. Finzi, F. Severi, G. Armellini, P. Caldirola, A. Aliotta; preface by Einstein and translation into Italian of his 1905, 1916, 1917, and 1953 memoirs on relativity theory.

F17 Petruccioli, Sandro, Carlo Tarsitani, Fabio Bevilacqua and Paswuale Tucci. Sulla Genesi Storica e sul Significato Teorico della Relativita di Einstein. Pisa: Domus Galilaeana, Quaderni di Storia e Critica della Scienza, Nuova Serie 4, (1973).

F18 Pyenson, L. "M. Plank-Redaktor 'Annalen der Physik': Bor'ba za utverzhdenie teorii otnositel'nosti." Voprosy Istorii Estestvoznaniia i Tekhniki, 1 (1982): 61-71.

Based upon the material of the correspondence between both editors of the Annalen der Physk, M. Planck and W. Wien, the article considers the editorial policy towards relativity theory. It deals with Planck's efforts in trying to achieve general recognition of Einstein's theory, paying special attention to the physical sense of the theory.

F19 Rossi, Arcangelo. "Aspetti storici ed epistemologici della critica operazionista delle theorie di A. Einstein." Physis, 21 (1979): 146-56.

On P. W. Bridgman's interpretation of relativity theory.

F20 Schmutzer, E. Relativitätstheorie-aktuell. Ein Beitrag zur Einheit der Physik. Mathematisch-Naturwissenschaftliche Bibliothek, Band 68, BSB. Leipzig: B.G.Teubner Verlagsgesellschaft, 1969. 180 pp.

Scientific and personal biography of Einstein.

F21 Sexl, R. "Schroedinger's Contribution to Relativity." The Schrödinger Equation. Edited by W. Thirring and P. Urban. (Acta Physica Austriaca, Supplement 17) New York: Springer-Verlag, 1977, pp. 7-18.

F22 Těšínká, Emilie. "K prijeti teorie relativity v ceske fyzice." Prace z dejin prirodnich ved, 13 (1980): 55-60.

F23 Toth, Imre. "Spekulationen über die Möglichkeit eins nicht euklidischen Raumes vor Einstein." Einstein Symposium, Berlin (item BE68), pp. 46-83.

F24 Whitrow, Gerard J. "The principle of relativity and the 'cosmic fireball'." Annali dell'Instituto e Museo di Storia della Scienza di Firenze, 5, fasc. 2 (1980): 83-91.

"The problem of the existence and relativity of inertial frames is considered in historical perspective and in the light of modern knowledge, concerning the structure of the universe, with particular reference to the discovery of the cosmic background black-body radiation."

F25 Whitrow, G. J. The Natural Philosophy of Time. 2nd ed. New York: Oxford University Press, 1980. 400 pp.

F26 Whittaker, Edmund. A History of the Theories of Aether and Electricity. 2 vols. London: Nelson, 1951, 1953. Reprint, New York: Humanities Press, 1973. xiv + 434 pp. and xi + 319 pp.

Volume I provides a comprehensive treatment of 19th century theories, Volume II covers both relativity and quantum theory and is notorious for the author's attempt to minimize the contribution of Einstein (Chapter II is titled "The Relativity Theory of Poincaré and Lorentz") but is otherwise a useful work.

F27 Williams, L. P., Ed. Relativity Theory: Its Origins and impact on modern thought. New York: Wiley, 1968. 164 pp.

Extracts from writings of Michelson & Morley, Lorentz, Poincare, Einstein, E. Whittaker, G. Holton, A. Gruenbaum, W. F. Magie, L. T. More, A. S. Eddington; "lay reaction" from New York Times (1928), H. W. Carr, H. Eliott, T. J. Graven, P. C. Squires, Jose Ortega y Gasset.

F28 Yezzi, Ronald D. The application of mathematics to concepts in physics: Four Theories. Ph. D. Dissertation, Southern Illinois University, 1969. 133 pp.

On the theories of Aristotle, Newton, Einstein and A. N. Whitehead. For summary see Dissertation Abstracts International, 29 (1969): 4061-A.

F29 Zahar, Elie. "Einstein, Meyerson and the role of mathematics in physical discovery." British Journal for the Philosophy of Science, 31 (1980): 1-43.

On E. Meyerson's principle of identity" and the development of the special and general theories of relativity.

* * * * * * * * * * * * * * * *

For additional references see HW, sections D.b, and F.a; also CB, section CB; biographies of Einstein.

FA. RELATIVITY THEORY
(Origins and development of special theory)

FA1 Bellone, Enrico. *La relatività da Faraday a Einstein*, Torino: Loescher, 1981. 243 pp.

Anthology of Fresnel, Stokes, Helmholtz, Riemann, Lobacheviskii, Faraday, Maxwell, Kelvin, Poincare, Lorentz and Einstein. Author claims that Einstein's relativity is not the outcome of a philosophical "crisis" of so-called classical physics but a revolutionary offspring of 19th century physics itself.

* Butler, J. W. "On the Trouton-Noble expepriment." Cited above as item E3.

FA2 Caldirola, Piero and Erasmo Recami. "Causality and Tachyons in Relativity." Dalle Chiara, Maria Luisa, Ed. *Italian Studies in the Philosophy of Science*. Boston: Reidel, 1980, pp. 249-298.

FA3 Cullwick, E. G. "Einstein and special relativity: Some inconsistencies in his electrodynamics." *British Journal for the Philosophy of Science*, 32 (1981): 157-76.

FA4 Frankfurt, U. I. "Optika dvizhushchikhsia sred i spetsial'naia teoriîa otnositel'nosti." *Ĕinshteĭnovskii Sbornik 1977*, pp. 257-326.

Optics of moving media and special relatitvity.

FA5 Galison, Peter Louis. "Minkowski's space-time: From visual thinking to the absolute world." *Historical Studies in the Physical Sciences*, 10 (1979): 85-121.

* Gruenbaum, A. "The bearing of philosophy on the history of science." Cited above as item M21.

FA6 Gutting, Gary. "Einstein's discovery of special relativity." *Philosophy of Science*, 39 (1972): 51-68.

FA7 Heerden, Van P. J. "On the history of the theory of relativity." *American Journal of Physics*, 36 (1968): 1171-72.

On the theory of H. A. Lorentz.

FA8 Holton, Gerald. "Resource Letter SRT-1 on Special Relativity Theory." *American Journal of Physics*, 30 (1962): 462-69.

Annotated bibliography.

FA9 Holton, Gerald. "Influences on Einstein's early work in relativity theory." <u>American Scholar</u>, 37 (1967/68): 59-79. Reprinted in <u>Thematic Origins</u> (item A46), pp. 197-217.

Discusses the influence of A. Foeppl.

FA10 Illy, Jozsef. "On the Birth of Minkowski's Four Dimensional World." <u>XIII International Congress of History of Science Proceedings, Moscow, 18-24 August 1971</u>. Section VI. Moskva: Nauka, 1974, pp. 67-72.

Suggests that Minkowski's apolegetic tone in his 1908 paper was due to the notoreity which 4-dimensional space had acquired through the spiritualist writings of the astronomer K. F. Zoellner.

FA11 Illy, Jozsef. "A speciális relativitáselmélet megszületese." <u>Különlenyomat a Fizikai Szemle</u> 25 (1975): 405-19

* Illy, Jozseph. "Lenin, elektromagneticky obraz světa a teorie relativity." Cited above as item FD8.

FA12 Illy, Jozsef. "Revolutions in a Revolution." <u>Studies in History and Philosophy of Science</u>, 12 (1981): 173-210.

Detailed survey of German literature on ether, electromagnetism and relativity, 1890-1920.

FA13 Kilmister, C. W. <u>Special Theory of Relativity</u>. New York: Pergamon Press, 1970. viii + 299 pp.

Includes extracts from works by A. A. Michelson, J. Larmor, H. A. Lorentz, A. Einstein, M. Wilson, H. A. Wilson, P. Zeeman, P. A. M. Dirac, C. D. Anderson, E. Wigner.

FA14 Knudsen, Ole. "19th Century Views on Induction in Moving Conductors." <u>Centaurus</u>, 24 (1980): 346-69.

The survey is intended to elaborate G. Holton's suggestion (item FA9) about the possible influence of A. Foeppl's discussion of this problem on Einstein.

FA15 Leplin, Jarrett. "The concept of an ad hoc hypothesis." <u>Studies in History and Philosophy of Science</u>, 5 (1975): 309-45.

On the Lorentz contraction hypothesis.

FA16 Lorentz, Hendrik Antoon, Albert Einstein, Hermann Minkowski. _Das Relativitaetsprinzip. Eine Sammlung von Abhandlungen_. Mit einem Beitrag von H. Weyl und Anmerkungen von A. Sommerfeld. 4th enl. ed. 160 pp. Leipzig: Teubner, 1922. Translation: _The principle of relativity: a collection of original memoirs on the special and general theory of relativity_, by H. A. Lorentz, A. Einstein, H. Minkowski, and H. Weyl. With notes by A. Sommerfeld. London: Methuen, 1923. Reprinted New York: Dover, 1952.

See items FA27 and FA29 for discussion of the translation of Einstein's paper.

FA17 Marder, Leslie. _Time and the Space-Traveller_. Philadelphia: University of Pennsylvania Press, 1971. 208 pp.

Semi-historical review of the "twin paradox" with bibliography of 305 items.

FA18 Mehra, Jagdish. "Albert Einstein's 'First' Paper." _Science Today_, (April 1971): 22-28.

English translation of unpublished manuscript (1894 or 1895) on the state of aether in magnetic fields. The German original is published by Mehra in _Physikalische Blaetter_, 27 (1971): 385-91.

FA19 Miller, Arthur I. _Albert Einstein's special theory of relativity: Emergence (1905) and early interpretation (1905-1911)_. Reading, Mass.: Addison-Wesley Pub.Co., 1981. xxviii + 466 pp.

A comprehensisve account which includes the work of H. A. Lorentz. H. Poincare, W. Kaufmann, M. Abraham; detailed analysis of Einstein's 1905 paper; further discussion by P. Ehrenfest, M. von Laue, and M. Planck. Extensive bibliography.

FA20 Miller, Arthur I. "A reply to 'Some new aspects of relativity: comments on Zahar's paper'." _British Journal for the Philosophy of Science_, 29 (1978): 252-56.

On the paper by Podlaha, item FA23.

FA21 Miller, Arthur I. "On the History of the Special Relativity Theory." _Albert Einstein: His Influence on Physics, Philosophy and Politics_ (item BE9), pp. 89-108.

An analysis of Einstein's 1905 paper in its historical context.

FA22 Ogawa, Tsuyoshi. "Japanese evidence for Einstein's knowledge of the Michelson-Morley experiment." *Japanese Studies in History of Science*, 18 (1979): 73-81.

J. Ishiwara's report of Einstein's 1922 lecture in Japan.

FA23 Podlaha, M. F. "Some new apects of relativity: Comments on Zahar's paper." *British Journal for the Philosophy of Science*, 27 (1976): 261-67.

Comments on item FA39, see also item FA23.

FA24 Prokhovnik, S. J. "Did Einstein's programme supercede Lorentz's?" *British Journal for the Philosophy of Science*, 25 (1974): 336-40.

Comments on the paper by Zahar, item FA39.

FA25 Reichel, Uwe. *Die Rolle der Experimente bei der Speziellen Relativitätstheorie*. Stuttgart: Historisches Institut, Zulassungsarbeit, 1970.

* Schaffner, Kenneth F. *Nineteenth-Century Ether Theories*. Cited below as item G38.

FA26 Schwartz, H. M. "Einstein's comprehensive 1907 essay on relativity." *American Journal of Physics*, 45 (1977): 512-17, 811-17, 899-902.

Includes translation of part of his article in *Jahrbuch der Radioaktivitaet*.

FA27 Schwartz, H. M. "Einstein's first paper on relativity." *American Journal of Physics*, 45 (1977): 18-25.

Includes annotated English translation of first part of paper.

FA28 Schwartz, H. M. "Poincaré's Rendiconti paper on relativity, Part I." *American Journal of Physics*, 39 (1971): 1287-94.

FA29 Scribner, Charles, Jr. "Mistranslation of a passage in Einstein's original paper on relativity." *American Journal of Physics*, 31 (1963): 398.

FA30 Seitz, Werner. *Arnold Sommerfelds Stellung zur Relativitäts - und Quantentheorie in den Jahren 1900-1911*. Stuttgart: Historisches Institut, Zulassungsarbeit, 1971.

FA31 Shankland, R. S. "Conversations with Albert Einstein, II." _American Journal of Physics_, 41 (1973): 895-901, 43 (1975): 464.

On the disagreement about when and how much Einstein knew about the Michelson-Morley experiment.

FA32 Shankland, R. S., et al. [Proceedings of the Michelson Colloquium in Potsdam, 28-29 April, 1981] _Astronomische Nachrichten_, 303 (1982): 1-96.

Includes papers by J. Auth, D. Michelson-Livingston, H. Melcher, R. Rompe & G. Albrecht, K. Lanius, H. G. Shopf, L. S. Swenson, Jr., J. Stachel, J. P. Vigier, Z. Maric, F. Kaschluhn, and J. J. Treder, mostly on the Michelson-Morley experiment and its relation to modern physics.

FA33 Starosel'skaîa-Nikitina, O. A. "Rol'Anri Puankare v sozdanii teorii otnositel'nosti." _Voprosy Istorii Estestvoznaniia i Tekhniki_, 5 (1957): 39-49.

H. Poincaré's role in the creation of relativity theory (with bibliography).

* Stein, Howard. "Subtler forms of matter." Cited below as item G42.

FA34 Swenson, Loyd S. _Genesis of Relativity: Einstein in Context_. New York: Franklin, 1979. 266 pp.

* Swenson, Loyd S. _The Ethereal Aether, a History of the Michelson-Morley-Miller Aether-Drift experiments, 1880-1930_. Cited below as item G44.

FA35 Swenson, L. S., Jr. "The Michelson-Morley-Miller experiments and the Einsteinian synthesis." _Astronomische Nachrichten_, 303 (1982): 39-45.

A personal account in which the author gives an overview of his historical research on the aether-drift experiments and their connection with relativity.

FA36 Tiapkin, A. A. "Ob Istorii Formirovaniîa Ideĭ Spetsial'noĭ Teorii Otnositel'nosti." _Printsip Otnositel'noti. Sbornik Rabot po Spetsial'noĭ Teorii Otnositel'nosti_. A. A. Tiapkin, Ed. Moskva, 1972, pp. 271-330.

On the history of formation of ideas of special relativity.

FA37 Toernebohm, Hakan. "Two studies concerning the Michelson-Morley experiment." _Foundations of Physics_, 1 (1970): 47-56.

Includes arguments on the relation between relativity theory and the existence of an ether.

FA38 Vizgin, V. P. _Reliativistskaia Teoriia Tiagoteniia: Istoki i formirovanie 1900-1915_. Moscow: Izdatel'stvo Nauka, 1981. 350 pp.

FA39 Zahar, Elie. "Why did Einstein's programme supersede Lorentz's?" _British Journal for the Philosophy of Science_, 24 (1973): 95-123, 223-62.

See critiques by P. K. Feyerabend, A. I. Miller and K. F. Schaffner, in _ibid_. 25 (1974): 25-78 and reply by Zahar, _ibid_. 29 (1978): 49-60.

* Zahar, Elie. "Crucial experiments: A Case Study." Cited above as item JB30.

* Zahar, Elie. "Mach, Einstein, and the Rise of Modern Science." Cited below as item K110.

* * * * *

For additional references see HW, section F.b.

FB. GENERAL RELATIVITY AND COSMOLOGY

FB1 Bergmann, Peter G. "The Development of the Theory of Relativity." *Albert Einstein: His influence on Physics, Philosophy and Politics* (item BE9), pp. 1-16.

FB2 Bičák, Jiří. "Einsteinova cesta k obecné teorii relativity." *Československý časopis pro fyziku*, sekce A, 29 (1979): 222-43.

FB3 Caldirola, Piero. "Contributo di Scienziati dell'Ateneo Pavese alla Teoria della Relatività." *Atti del 4 Convegno Nazionale di Relatività e Fisica della Gravitazione, Collegio Ghilieri, Pavia, 23 - 26 settembre 1980*. Bologna: Technoprint, 1981, pp. 1-4.

Survey of the contributions of Eugenio Beltrami (1835-1899), Carlo Somigliana, Attilio Palatini and Rocco Serini to the development of non-euclidean geometry and general relativity.

FB4 Chandrasekhar, S. "The General Theory of Relativity: The First Thirty Years." *Contemporary Physics*, 21 (1980): 429-49.

The author's "impression of the principal landmarks."

FB5 Chandrasekhar, S. "The role of General Relativity in Astronomy: Retrospect and Prospect." *Journal of Astrophysics and Astronomy*, 1 (1980): 33-45.

FB6 Chandrasekhar, S. "The increasing role of general relativity in astronomy." *Observatory*, 92 (1972): 160-74.

FB7 Chandrasekhar, S. "The Richtmyer Memorial Lecture -- Some Historical Notes." *American Journal of Physics*, 37 (1969): 577-84.

On Eddington's role in the 1919 eclipse expedition.

FB8 Dorling, Jon. "Did Einstein need General Relativity to solve the problem of absolute space? Or had the problem already been solved by Special Relativity?" *British Journal for the Philosophy of Science*, 29 (1978): 311-23.

FB9 Earman, John and Clark Glymour. "The Gravitational Red Shift as a Test of General Relativity: History and Analysis." *Studies in History and Philosophy of Science*, 11 (1980): 175-214.

FB10 Ferraris, M., M. Francaviglia and C. Reina. "Variational Formulation of General Relativity from 1915 to 1925 "Palatini's Method" discovered by Einstein in 1925." *General Relativity and Gravitation*, 14 (1982): 243-254.

"The metric-affine variation principle, according to which the metric and affine connection are varied independently, is commonly known as the "Palatini method." In this paper, we revisit the history of the "golden age" of general relativity, through a discussion of the papers involving a variational formulation of the field problem. In particular we find that the original Palatini paper of 1919 was rather farfrom what is usually meant by 'Palatini's method,' which was instead formulated, to our knowledge, by Einstein in 1925." (Author's abstract)

FB11 [Friedman, A. A.] "Pamiati A. A. Fridmana - k 75-letiiu so dnia rozhdeni͡a." *Uspekhi Fizicheskikh Nauk*, 800, no. 3 (1980).

Contains two classic papers of A. A. F. on relativistic cosmology, Einstein's observations, articles by V. A. Fok, P. Ia. Pulubarinova-Kochina, Ia. B. Zel'dovich, E. M. Lifshits, I. M. Khalatnikov on Friedman and the genesis and development of relativistic cosmology to the beginning of the sixties.

FB12 Frolov, V. P. "Fizika chernykh dyr: ot Eĭnshteĭna do nashikh dneĭ." *Eĭnshteĭnovskiĭ Sbornik 1975-76*. Moskva: Nauka, 1978, pp. 82-151.

Black hole physics from Einstein to the present.

FB13 Guth, E. "Contribution to the history of Einstein's geometry as a branch of physics." *Proceedings of the Relativity Conference in the Midwest, 1969*. Edited by Mosche Carmeli et al. New York: Plenum Press, 1970, pp. 161-207.

FB14 Held, A., Ed. *General Relativity and Gravitation, One Hundred Years after the Birth of Albert Einstein*. New York: Plenum, 1980. 2 vols. xviii + 540 pp.; xx + 598 pp.

FB15 Heller, M., and O. Godart. "Origins of relativistic cosmology." *Astronomy Quarterly*, 4, no. 13 (1981): 27-33.

On the theories of Einstein, de Sitter and Lemaitre.

FB16 Hetherington, Norriss S. "Sirius B and the Gravitational Redshift: An historical review." *Quarterly Journal of the Royal Astronomical Society*, 21 (1980): 246-52.

FB17 Hoffmann, Banesh. "Einstein and tensors." Tensor, n.s. 26 (1972): 157-62.

FB18 Ivanenko, D. D. "Neprekhodiâshchaîa Aktual'nost Teorii Gravitatsii Éĭnshteĭna." Voprosy Istorii Estestvoznaniiâ i Tekhniki, 67-68 (1979): 3-14.

On Einstein's general relativity.

FB19 Kilmister, G. W. General Theory of Relativity. New York: Pergamon Press, 1973. ix + 365 pp.

Includes extracts from works of B. Riemann, W. K. Clifford, A. Einstein, L. Infeld, V. Fock, F. A. E. Pirani, H. Bondi, M. G. J. van der Burg, A. W. K. Metzner, J. R. Oppenheimer, H. Snyder, R. Penrose, R. V. Pound and G. A. Rebka, Jr.

FB20 Magnani, L., ed. Le geometrie non euclidee. Bologna: Zanichelli, 1979. iv + 182 pp.

Anthology of Euclid, Kant, Riemann, Beltrami, Poincaré, Bachelard and Nagel.

FB21 Masani, Alberto. Storia della cosmologia. Roma: Editori Riuniti, 1980. 267 pp.

FB22 Mehra, J. "Einstein, Hilbert, and the theory of gravitation." The Physicist's Conception of Nature (item A80), pp. 92-178. Also published as a separate monograph: Boston: Reidel, 1974. 88 pp.

FB23 Parrini, Paolo. Fisica e geometria dall'Ottocento ad oggi. Tornio: Loescher, 1979. 252 pp.

FB24 Perlmutter, Arnold and Linda F. Scott, eds. On the Path of Albert Einstein. New York: Plenum Press, 1979. vii + 177 pp.

Part of the Proceedings of Orbis Scientiae, 1979, held by the Center for Theoretical Studies, University of Miami, Coral Gables, Florida, January 15-18, 1979. Preface by the Chairman, Behram Kursunoglu. Includes: B. Kursunoglu, "A non-technical history of the generalized theory of gravitation;" D. F. Moyer, "Revolution in science: The 1919 eclipse test of general relativity."

* Pyenson, Lewis. "Mathematics, Education, and the Göttingen Approach to Physical Reality, 1890-1914." Cited above as item C90.

* Reines, Frederick, Ed. Cosmology, Fusion & Other Matters. Cited above as item BG3.

FB25 Sewell, William C. Einstein, Mach, and the General Theory of Relativity. Ph.D. Dissertation, Case Western Reserve University, 1975. 324 pp.

For summary see Dissertation Abstracts International, 36 (1975): 3774A.

FB26 Stein, Howard. "Some philosophical prehistory of General Relativity." Foundations of Space-Time Theories (Minnesota Studies in the Philosophy of Science, Volume VIII). Edited by John Earman, Clark Glymour, and John Stachel. Minneapolis: University of Minnesota Press, 1977, pp. 3-49.

FB27 Tauber, Gerald E., Ed. Albert Einstein's Theory of General Relativity. New York: Crown Publishers, 1979. 352 pp.

Consists mainly of reprints of previously published material and short essays reviewing the currrent status of various aspects of general relativity, its experimental verification and astrophysical implications. Of historical interest are the recollections by J. A. Wheeler, M. Wellner, A. Komar and O. W. Greenberg of their May 17, 1953 visit to Einstein; G. Holton, "On Einstein's Weltbild;" and J. L. Synge's recollections of "My relativistic milestones."

FB28 Treder, H. J. "Einstein und die Potsdamer Astronomen." Die Sterne, 55 (1979): 1-10.

On K. Schwarzschild, E. F. Freundlich, etc.

FB29 Vizgin, V. P. "'Erlangenskiĭ' Podkhod k Istorii Fiziki." Trudy Komissii͡a po Istorii Estestvoznanii͡a i Tekhniki AN Lit SSR, X-Xi, Sektsii͡a Istorii Fiziki (1968): 44-50.

The "Erlangen" path to history of physics.

FB30 Vizgin, Vladimir. "On the Road to the Relativity Theory of Gravitation." Soviet Studies in the History of Science. Moscow: Social Sciences Today Editorial Board, 1977, pp. 135-46.

FB31 Vizgin, V. P. "Eksperiment i Obshchai͡a Teorii͡a Otnositel'nosti." Voprosy Istorii Estestvoznanii͡a i Tekhniki, 67-68 (1980): 22-33.

Experiment and general relativity theory.

* * * * * * * * * * *

For additional references see HW, sections F.d and P.k

FC. GRAVITY

FC1 Amaldi, E. "Einstein and Gravitational Radiation." _Gravitation Radiation, Collapsed Objects and Exact Solutions_. Edited by C. Edwards. Berlin: Spring-Verlag, 1980, pp. 246-98.

Begins with a brief historical account of Einstein's mathematical discovery.

FC2 Apolin, A. "Die Geschichte der Gravitation." _Philosophia Naturalis_, 12 (1970): 156-72.

FC3 Bertotti, Bruno. "Scienza normale e scienza di crisi nella fisica della gravitazione." _Teoria fisica e realtà_ (item A18), pp. 113-27.

FC4 Collins, H.M. "The seven sexes: A study in the sociology of a phenomenon, or the replication of experiments in physics." _Sociology_, 9 (1975): 205-24.

On J. Weber's claim to have discovered gravity waves.

FC5 Collins, H. M. "Son of seven sexes: the social destruction of a physical phenomenon." _Social Studies of Science_, 11 (1981): 33-62.

Recent developments concerning J. Weber's claim to have discovered high-flux gravitational radiation; shows the "social processes involved in the growth of almost universal disbelief in the positive claims that were being made up to the mid-1970s."

FC6 Everitt, C. W. F. "Gravitation, relativity and precise experimentation." _Proceedings of the First Marcel Grossmann Meeting on General Relativity_. Edited by R. Ruffini. New York: North-Holland, 1977, pp. 545-615.

On the experiments of C. V. Boys (1889), Michelson & Morley (1881, 1887), R. V. Eotvos (1890-1922), etc. and the current gyro relativity satellite experiment; attempts to detect gravitation waves by J. Weber and others.

FC7 Gillies, George T. _The Newtonian Gravitational Constant: An Index of Measurements_. Sevres, France: Bureau International des Poids et Mesures, 1982. 83 pp.

FC8 Ivanenko, D. "60 Years of Gravitational Physics in the USSR." _Papers by Soviet Scientists_ (item M48), pp. 11- 28.

FC9 Ivanenko, D. D. "50 Let Sovetskikh Rabot Po Gravitatsii." *Izvestiia Vysshikh Uchebnykh Zavedenii*. Fizika, 10 (1967): 30-38.

50 years of Soviet work in gravitation.

FC10 Newman, Ezra T. "This Week's Citation Classic: Newman E. and R. Penrose. An approach to gravitational radiation by a method of spin coefficients. *J. Math.Phys*. 3: 566-78, 1962." *Current Contents, Physical, Chemical & Earth Sciences*, 21, no. 19 (May 11, 1981): 20.

The author recalls the circumstances of writing this paper, which has been cited more than 425 times since 1962.

FC11 Vizgin, V. P. "Teorii͡a T͡iagoteni͡a na Rubezhe XIX i XX vv." *Eĭnshteĭnovskii Sbornik 1975-76*. Moskva: Nauka, 1978, pp. 245-282.

Theory of gravitation in the 19th & 20th centuries.

FC12 Whitrow, G. J. "From the problem of fall to the problem of collapse: Three hundred years of gravitational theory." *Philosophical Journal*, 14 (1977): 67-84.

FC13 Zemplen, Jolan M. "The Eotvos experiment and modern physics." *Acte, XIIe Congres International d'Histoire des Sciences*, 5 (1968, pub. 1971): 121-25.

FD. RELATIVITY: POPULAR RESPONSE, CULTURAL INFLUENCES, ETC.

FD1 Biezunski, Michel. "Einstein à Paris." *La Recherche*, 132 (April, 1982): 502-10.

Deals with reactions of Paris press and public to Einstein's 1922 lectures on relativity.

FD2 Cohen, M. R. "Einstein in the Movies -- A note on the exposition in the films of the theory of relativity." *Vanity Fair*, 20, no. 6 (Aug. 1923): 48, 96.

FD3 Creed, Walter G. *Contemporary scientific concepts and the structure of Lawrence Durrell's Alexandria Quartet*. Ph.D. Dissertation, University of Pennsylvania, 1968. 176 pp.

For summary see *Dissertation Abstracts International*, 30 (1969): 1165-A

FD4 Delokarov, K. Kh. *Bor'ba Sovetskikh Filosofov i Fisikov Za Dialektiko-Materialisticheskoe Istolkovanie Teorii Otnositel'nosti i Kvantovoĭ Mekshaniki v SSSR. 1920-1930gg*. Avtoref. Dis. Na Soisk, Uch. Step. Kand. Filos. Nauk. Moskva, 1968. 21 pp. (Moskovskiĭ Gosudarstvennyĭ Pedagogicheskiĭ Institut Im. V.I. Lenina).

The battle of Soviet physicists and philosophers for a dialectico-materialistic interpretation of relativity and quantum mechanics: 1920-30.

FD5 Delokarov, K. Kh. "Bor'ba Za Materialiticheskoe Istolkovanie Teorii Otnositel'nosti v Sovetskoĭ Nauke. 1920-1930 gg." *Nauchnye Doklady Vyssheĭ Shkoly, Filos*. Nauki, 4 (1967): 125-132.

The battle for a materialistic interpretation of relativity in Soviet Science: 1920-30.

FD6 Giacomini, Ugo. *Spazio e tempo nel pensiero contemporaneo*. Genova: Commissionaria Editrice Tilgher, 1975. 246 pp.

Philosophical problems of relativity, etc.

FD7 Hermann, A. "Tendenzen der zwanziger Jahre. Der Kampf um die Relativitaetstheorie." *Bild der Wissenschaft*, 9 (1977): 108-116.

FD8 Illy, Jozsef. "Lenin, elektromagneticky obraz světa a teorie relativity." *Revoluční změny v oblasti vědy a techniky. Prace z. dějin přirodnih věd*, no. 13, (1980): 31-38.

"Lenin, the electromagnetic world view, and the theory of relativity."

FD9 Mechling, Jay. "In Search of an American Ethnophysics." The Study of American Culture: Contemporary Conflicts. Edited by Luther S. Luedtke. Deland Florida: Everett/Edwards, 1977, pp. 241-277.

The American Studies community is in the stage of "transition to maturity" in a Kuhnian paradigm revolution. The old myth/symbol/image paradigm is being replaced by "cognitive anthropology" or "new ethnography". Does relativity theory define a specific community of American physicists in the early 20th century? The author discusses whether it is possible to discuss this question from the viewpoint of his hypothetical ethnophysics.

* Nadeau, Robert. Readings from the new book on Nature: Physics and Metaphysics in the Modern Novel. Cited below as item L23.

FD10 Ringold, Francine. "The Metaphysics of Yoknapatawpha County: 'Airy Space and Scope for your Delirium.'" Hartford Studies in Literature, 8 (1976): 223-40.

Influence of space-time theories of Einstein and Bergson on Faulkner's Absalom, Absalom!

FD11 Sachs, Mendel. "On the philosophy of general relativity theory and ideas of Eastern and Western cultures." The Ta-You Wu Festschrift. Edited by S. Fujita. New York: Gordon & Breach, 1978, pp. 9-24.

FD12 Scheick, William J. "The Fourth Dimension in Wells's Novels of the 1920s." Criticism, 20 (1978): 167-90.

"The subject of the interaction between space and time, especially as emphasized in Albert Einstein's Special Theory of Relativity, bears upon Wells's fiction of the Twenties for Wells's aim then was to create a fictional 'space' or form which undergoes duration, is synchronically 'timely', is relative to the reader."

FD13 Torrance, Thomas F. "Newton, Einstein and scientific theology." Religious Studies, 8 (1972): 233-50.

On the views of Karl Barth compared to those of Einstein.

FD14 Troy, William. "Time and Space Conceptions in Modern Literature." Selected Essays. Edited by S. E. Hyman. New Brunswick, NJ: Rutgers University Press, 1967. 1934, pp. 19-34.

FD15 Ziolkowski, Theodore. "Hermann Broch and Relativity in Fiction." Wisconsin Studies in Contemporary Literature, 8 (1967): 365-76.

G. OPTICS AND ELECTROMAGNETIC WAVES

G1 Bloembergen, Nicolaas. "Nonlinear Optics and Spectroscopy." Science, 216 (1982): 1057-64.

Nobel Lecture, 8 December 1981.

G2 Blok, G., "Magicheskiĭ Luch." Kul'tura i Zhizn, 9 (1963): 39-41.

History of laser light based on work of Soviet researchers V. Fabrikant, F. Butaev and M. Budynski.

G3 Bolotovskiĭ, B. M. Svechenie Vavilova-Cherenkova. Moskva: Nauka, 1964. 95 pp.

Introduction, chapter 1 contains a history of the discovery of the Vavilov-Cerenkov effect, 3-24.

G4 Braun, Carl Ferdinand. "Electrical oscillations and wireless telegraphy." Nobel Lectures...Physics (item A87), Vol. I, pp. 226-45. December 11, 1909.

G5 Carruthers, P. "Resource Letter GSL-1. Quantum and Statistical Aspects of Light." American Journal of Physics, 31 (1963): 321-25.

Annotated bibliography. Available also in booklet, Quantum and Statistical Aspects of Light, with selected reprints of papers by E. O. Lawrence and J. W. Beams (1928), E. M. Purcell (1956), W. E. Lamb, Jr. and R. C. Retherford (1947), N. Bohr and L. Rosenfeld (1950), R. H. Dicke (1954), R. Hanbury Brown and R. Q. Twiss (1956) and others published by the American Institute of Physics, New York (n.d.)

G6 Cherenkov, P. A., I. E. Tamm, I. M. Frank. Nobelevskie Lektsii. Moska: Fizmatgiz, 1960. 75 pp.

Nobel lectures of 1958 Nobel laureates Cherenkov, Tamm, Frank.

G7 Čerenkov, Pavel A. "Radiation of particles moving at a velocity exceeding that of light, and some of the possibilities for their use in experimental physics." Nobel Lectures...Physics (item A87),vol. 3, pp. 446-40. December 11, 1958.

G8 Danilychena, M. N. "Liuminestsentsiia kak neravnovesnoe izluchenie (k istorii voprosa)." Voprosy Istorii Estestvoznaniia i Tekhniki, no. 1 (54) (1976): 49-52.

History of the problem of luminescence as a non-equilibrium radiation. It is shown that research on luminescence played a part in the foundation of quantum mechanics; in some cases it was associated with the study of the dependence of optical parameters on the intensity of light (non-linear optics).

G9 (Denisiuk, Iu. N.) "Golografiia: Vozniknovenie i razvitie (Interviû c chlenom-korrespondentom AN SSSR Iu. N. Denisiûk)." Voprosy Istorii Estestvoznaniia i Tekhniki, 4 (1981): 104-10.

Interview by V.A. Gurikov with Iurii Nikolaevvich Denisiuk on the origin and development of holography.

G10 Ewald, P. P. "A Review of my papers on crystal optics 1912 to 1968." Acta Crystallogr., Sect. A, A35, Pt. 1 (1 Jan. 1979): 1-9.

G11 Fabelinskii, I. L. "The discovery of combinational scattering of light (the Raman effect)." Soviet Physics Uspekhi, 21 (1978): 780-97. Translated from Uspekhi Fizicheskikh Nauk, 126 (1978): 124-52.

Original papers by A. V. Raman, K. S. Krishnan, L. S. Landsberg, and L. I. Mandel'shtam. Brief outline of Soviet research on this subject using archival materials.

G12 Fabrikant, V. A. "40 Let Sovetskoĭ Fizicheskoĭ Optiki." Svetotekhnika, 11 (1957): 3-9.

40 years 1917-1957 of Soviet physical optics.

G13 Frank, Il'ja M. "Optics of light sources moving in refractive media." Nobel Lectures...Physics (item A87), vol. 3, pp. 442-68. December 11, 1958.

G14 Franken, Peter. "Optics: an ebullient evolution." Physics Today, 34, no. 11 (Nov. 1981): 160-71.

Survey of recent developments, especially the laser.

G15 Genzel, Ludwig and K. Sakai. "Interferometry from 1950 to the present." Journal of the Optical Society of America, 67 (1977): 871-79.

G16 Gerasimov, F. M. "Razvitie Rabot Po Difraktsionnym Reshetkam v GOI." Trudy Gosudarstvennyi Opticheskii Institut im S. I. Vavilova, 29, 158 (1965): 64-83.

On the development of diffraction gratings at the State Optical Institute (Moscow).

G17 Ginsburg, N. "History of Far-Infrared Research II. The Grating Era, 1925-1960." Journal of the Optical Society of America, 67 (1977): 865-71.

Traces developments of far infra-red spectroscopy and discusses increase in spectral range and improvement in resolution resulting from the use of semiconductor devices.

G18 Gvozdover, S. D. "Razvitie Radiofiziki i Elektroniki Na Fizicheskom Fakul'tete MGU." Istoriia i Metodologiia Estestvennykh Nauk, 2 (1963): 247-63.

Development of radiophysics and electronics at the Physics Dept. of Moscow State University.

G19 Ivankov, A. G. "O Vozniknovenii Radiospektroskopii." Istoriia i Metodologiia Estestvennykh Nauk, 4 (1966): 231-36.

The beginnings of radiospectroscopy in Russia at the turn of the 19th-20th centuries.

* (Kingslake, Hilda G. et al.) "History of the Optical Society of America, 1916-1966." Cited above as item C61.

G20 Konno, Hiroyuki. "Some remarks on the interference experiments with faint light." Historia Scientiarum, 20 (1981): 95-105.

On the experiments of G. I. Taylor (1909) as related to J. J. Thomson's ideas (1907) and on the experiments of A. J. Dempster and H. F. Batho (1927). Although such experiments were discussed during the development of quantum mechanics, they did not have a direct influence at the time.

G21 Korolev, F. A. "Optika i spektroskopiia." Razvitie Fiziki v Rossii (item A93), pp. 25-52.

G22 Kravets, T. P. "30 Let Sovetskoĭ Optiki." (1947). Kravets, T. P. Ot N'iutona Do Vavilova Ocherki i Vospominaniia, Leningrad (1967), pp. 244-273.

Thirty years of Soviet optics (1947).

G23 Krishnan, R. S. and R. K. Shankar. "Raman effect: history of the discovery." Journal of Raman Spectroscopy, 10 (Jan. 1981): 1-8.

Stresses the contribution of Raman and his students and also touches upon work of French and Russian groups.

G24 Levshin, V. L. "Vklad Sovetskoĭ Nauki v Izuchenie Liuminestsentsii." Optika i Spektroskopiia, T.3, 5 (1957): 417-433.

Survey of 40 years of Soviet contributions to

luminescence studies.

G25 Levshin, L. V. "Liuminestentsiia." Razvitie Fizike v Rossii (item A93), pp. 52-76.

G26 Lippmann, Gabriel. "Colour photography." Nobel Lectures...Physics (item A87), Vol. I, pp.186-88. December 14, 1908.

G27 Loewenstein, Ernest V. "The History and Current Status of Fourier Transform Spectroscopy." Applied Optics, 5 (1966): 845-854.

G28 Lorentz, H. A. "The theory of electrons and the propagation of light." Nobel Lectures...Physics (item A87), Vol. I, pp. 14-29. December 11, 1902.

G29 Lubkin, G. B. "Nobel Physics Prize to Bloembergen, Schawlow, and Siegbahn." Physics Today, 34, no. 12 (Dec. 1981): 17-20.

"The 1981 Nobel Prize in physics has been awarded to Nicolaas Bloembergen of Harvard University, Arthur L. Schawlow of Stanford University and Kai M. Siegbahn of Uppsala University. Bloembergen and Schawlow will receive half the prize 'for their contribution to the development of laser spectroscopy'. The other half will go to Siegbahn 'for his contribution to the development of high-resolution electron spectroscopy'." The article presents a brief history of the research for which the prize is awarded.

G30 Mandel, Leonard and Emil Wolf. "This week's citation classic. Mandel L & Wolf E. Coherence properties of optical fields. Rev. Mod. Phys. 37: 231-87, 1965."Current Contents, Physical, Chemical & Earth Sciences, 20, no. 47 (24 Nov. 1980): 16.

The authors recall the circumstances under which they wrote this paper, which has been cited over 435 times since 1965.

G31 Marconi, Guglielmo. "Wireless telegraphic communication." Nobel Lectures...Physics (item A87), Vol. I, pp. 196-222. December 11, 1909.

G32 Marshall, Eliot. "Gould advances inventor's claim on the laser." Science, 216 (1982): 392-95.

On March 1, 1982, a federal judge in San Francisco ruled that a patent filed by Gordon Gould in 1959 and awarded in 1977 is valid. See below, item G46. For an update on the case see Eliot Marshall, "Laser Wars in Court: Gould v. Bell Telephone," ibid. 217 (1982): 810.

G33 Michelson, Albert A. "Recent advances in spectroscopy." *Nobel Lectures...Physics*, (item A87), Vol. I, pp. 166-78 . December 12, 1907.

G34 "1917-1957 gg." *Radiotekhnika i Elektronika*, 11 (1957): 1319-1343.

Development of radiotechnology and electronics in USSR. Bibliography, pp. 1340-3.

* Nye, Mary Jo. "Gustave LeBon's Black Light: A study in physics and philosophy in France at the turn of the century." Cited below as item K79.

G35 Palik, E. D. "History of Far-Infrared Research I. The Rubens Era." *Journal of the Optical Society of America*, 67 (1977): 857-65.

Review article covering the period between 1892 and 1925 during which the work of H. Rubens progressed from near to far infrared.

G36 Parker, M. R. "The Kerr-Magneto-Optic Effect. (1876-1976)." *Physica*, 86-88B (1977): 1171-1176.

G37 Pease, Paul L. "Resource Letter CCV-1: Color and color vision." *American Journal of Physics*, 48 (1980): 907-17.

Annotated bibliography.

G38 Schaffner, Kenneth F. *Nineteenth-Century Aether Theories*. Oxford: Pergamon Press, 1972. ix + 278 p.

Includes a discussion of the electromagnetic theory of H. A. Lorentz, and extracts from his works of 1895 and 1899.

G39 Schawlow, Arthur L. "Spectroscopy in a New Light." *Science*, 217 (1982): 9-16.

Nobel Lecture, 8 December, 1981.

G40 Shushurin, S. F. "K istorii golografii." *Uspekhi Fizicheskikh Nauk*, 105 (1971): 145-48.

G41 Siphorov, V. I. "Radioelektronika v SSSR." *Kul'tura i Zhizn'*, 3 (1959): 31-34.

Survey of development of radioelectronics from 1895.

G42 Stein, Howard. "Subtler forms of matter." *Conceptions of Ether: Studies in the History of Ether Theories 1740-1900*. Edited by G. N. Cantor and M. J. S. Hodge.

New York: Cambridge University Press, 1981, pp. 309-40.

On the ether/electromagnetic theories of J. MacCullagh, G.F. Fitzgerald, H.A. Lorentz, Lord Kelvin, J. Larmor, etc.

* Stoicheff, Boris P. "Laser spectroscopy shares prize." Cited above as item JE4.

G43 Svelto, Orazio. "The laser: Physical principles and historical review." Europhysics News, 12, no. 7 (July 1981): 1-3.

G44 Swenson, Loyd S. The Ethereal Aether. A History of the Michelson-Morley-Miller Aether-Drift Experiments, 1880-1930. Austin: U. of Texas Press, 1972. xxii + 361 pp.

A survey of attempts to measure the earth's motion in relation to the "ether" concept, including the experiments of D. C. Miller in the 1920s as well as the more famous Michelson-Morley experiment.

G45 Tamm, Igor E. "General characteristics of radiations emitted by systems moving with super-light velocities with some applications to plasma physics." Nobel Lectures..Physics (item A87), vol. 3, pp. 470-82. December 11, 1958.

G46 Trotter, Robert J. "Laureates hit ruling on invention of laser." Chronicle of Higher Education, 24, no. 3 (17 March 1982): 3.

C. H. Townes and A. L. Schawlow, who won the Nobel Prize for inventing the laser, criticized the decision by a U.S. district court in San Francisco that Gordon Gould's laser patent is valid (see item G32 above).

G47 Worrall, John. "The pressure of light: The strange Case of the Vacillating 'Crucial Experiment'." Studies in History and Philosophy of Science, 13 (1982): 133-71.

Includes the experiments of P. Lebedew (1902), E. F. Nichols & G. F. Hull (1904), and the critique of the latter by Mary Bell and S. E. Green (1933, 1934).

G48 Zeeman, Pieter. "Light radiation in a magnetic field." Nobel Lectures...Physics (item A87), vol. I, pp. 33-40. May 2, 1903.

G49 Zernike, Frits. "How I discovered phase contrast." Nobel Lectures...Physics (item A87), vol. 3, pp. 239-46. December 11, 1953.

* * * *

For additional references see HW, section G.

H. QUANTUM THEORY (General and Miscellaneous)

H1 Abro, A. d'. The Rise of the New Physics: Its Mathematical and Physical Theories. New York: Dover Publications, 1951. (reprint of Decline of Mechanism, first published in 1939). 982 pp.

Volume II, Quantum Theory, provides a comprehensive account of the technical and conceptual developments, but without any references to the literature or evidence of detailed historical research by the author.

H2 Agassi, Joseph. "Quanta in context." Einstein Symposium, Berlin (item BE68), pp. 180-203.

H3 Andrade e Silva, J. L. and Georges Lochak. Quanta (Translated from the French by Patrick Moore) New York: McGraw-Hill, 1959. 248 pp.

A readable account of the development of quantum theory, suitable for students; not completely up-to-date from the standpoint of historical research though it contains a good selective bibliography.

H4 Ariyama, Kanetaka. "50 Years of the Quantum Theory. I. Theory of the Assembly of Molecular Systems" (in Japanese). Journal of History of Science, Japan, no. 21 (1951): 1-4.

H5 Bevilacqua, Fabio. "Albert Einstein e lo sviluppo della meccanica quantistica." Contributi (item H10), pp. 103-18.

* Bohr, Niels. Essays 1958-1962 on Atomic Physics and Human Knowledge. Cited below as item K9.

H6 Brauer, Wolfram; Hans-Waldemar Streitwolf, and Kurt Werner, eds. 75 Jahre Quantentheorie. Festband zum 75. Jahrestag der Entdeckung der Planckschen Energiequanten. Berlin: Akademie-Verlag, 1977. 306 pp.

Includes R. Rompe, "Zur Geschichte der Beziehungen der Quantenphysik und der Technik in den ersten Jahrzehnten des 20. Jahrhunderts"; F. Hund, "Das Korrespondenzprinzip als Leitfaden zur Quantenmechanik von 1925," B. Pontecorvo, "Die Neutrinophysik - gestern und heute"; W. Brauer, "Die Physik der Metalle und ihre Beeinflussung durch die Quantentheorie"; J. Auth, "Quantentheorie und Halbleitertechik."

H7 Brush, Stephen G. "Irreversibility and Indeterminism: Fourier to Heisenberg." _Journal of the History of Ideas_, 37 (1976): 603-30.

Argues that indeterminism was not suddenly introduced in quantum mechanics as a sharp break from classical physics, but emerged as the result of a trend started in the 19th century.

H8 Carazza, Bruno and Gian Paolo Guidetti. "Bohr e il principio di Corrispondenza." _Seminario di Storia della Fisica_ (item A110): 97-112.

H9 Center for History of Science and Technology, The Bancroft Library, University of California at Berkeley. _Inventory of Additions to the Archive for History of Quantum Physics since 1964_. Berkeley, 1973.

See item H22 for the original collection.

* Cline, Barbara Lovett. _Men who made a new physics_ (cited above as item B7).

H10 _Contributi alla storia della meccanica quantistica_. Pisa: Domus Galilaeana, Quaderni di Storia e Critica della Scienza, Nuova Serie, 7 (1976).

Contains items H5, H39, HA6, HB2, HB4, HC17.

H11 Dietz, David. _The Story of Science_. New York: The New Home Library, 1943. 387pp.

Part III deals with "The Story of the Atom" with a narrative on the recent development of physics.

H12 Elyashevich, M. A. "Fifty Years of Quantum Mechanics Discovery." _Papers by Soviet Scientists_ (item M48), pp. 1-10.

* Frenkel, Victor Ya. "Iz istorii prilozheniia kvantovomekhanicheskikh predstavlenii k teorii svobodnykh elektronov." Cited below as item DG22.

H13 Gamow, George. _Thirty Years that Shook Physics: The Story of Quantum Theory_. New York: Doubleday, 19676. xiv + 224 pp. Translation: _Trent'anni che sconvolsero la fisica. La storia della Teoria dei Quanti_. Bologna: Zanichelli, 1980. 205pp.

This book is valuable primarily for the author's personal viewpoint, lively style, and illustrations.

H14 Guillemin, Victor. The Story of Quantum Mechanics. New York: Scribner's, 1968. xvi + 333 pp.

A useful textbook; includes sections on elementary particles and philosophical aspects.

H15 Haar, D. ter. The Old Quantum Theory New York: Pergamon Press, 1967. x + 206 pp.

Includes translations or reprints of original papers by M. Planck (1900), A. Einstein (1905, 1917), E. Rutherford (1911), N. Bohr (1913), J. Franck and G. Hertz (1914) and W. Pauli (1925).

H16 Hermann, Armin and Ulrich Benz. "Quanten- und Relativitatstheorie im Spiegel der Naturforscherversammlungen" 1906-1920. Wege der Naturforschung 1822-1972 im Spiegel der Versammlungen Deutscher Naturforscher und Artze. Edited by H. Querner and H. Schipperges. Berlin, 1972, pp. 125-37.

H17 Hertz, G. "Aus den Anfangsjahren der Quantenphysik." Sitzungsberichte, Akademie der Wissenschaften DDR, Math.-Naturwiss.-Techn., Jg. 1975, nr. 15, 17 (1976). Russian translation in Uspekhi Fizicheskikh Nauk, 122 (1977): 497-511.

H18 Hund, Friedrich. "Paths to quantum theory historically viewed." Physics Today, 19, no. 8 (August 1966): 23-29. Translation: "Hatte die Geschichte der Quantentheorie auch anders ablaufen konnen?" Physikalische Blatter, 31 (1975): 29-35.

H19 Hurt, C. D. "A test of differences in the literature history of four historical accounts of the quantum mechanics problem." Scientometrics, 3 (1981): 457-66.

Using books by Hund, M. Jammer, and B. L. Van der Waerden and the inventory of the Quantum Physics Archive by T. Kuhn et al., the author concludes that "the literature history of quantum mechanics, when plotted as a function of frequency of publication date, is non-normal, negatively skewed, and is platykurtic."

H20 Jammer, Max. The Conceptual Development of Quantum Mechanics. New York: McGraw-Hill, 1966. xii + 399 pp.

One of the first works to use the Quantum Physics Archive (item H22), this is a thorough treatise on the history of quantum theory with extensive references to original sources.

H21 Jammer, Max. "Albert Einstein und das Quantenproblem." Einstein Symposium, Berlin (item BE68), pp. 146-67.

H22 Kuhn, Thomas S., John L. Heilbron, Paul Forman, and Lini Allen. _Sources for History of Quantum Physics. An Inventory and Report_. Philadelphia: American Philosophical Society, 1967 (its _Memoirs_, volume 68). ix + 176 pp.

Preface by J. A. Wheeler. Consists primarily of a catalog of letters, manuscripts and interviews; copies of the source materials were deposited initially at Berkeley, Copenhagen, and Philadelphia. See also item H9.

H23 Maiocchi, Roberto. "Paul Langevin e la meccanica quantistica." _Seminario di Storia della Fisica_ (item A110), pp. 167-198.

* Mehra, Jagdish. _The Solvay Conference on Physics_. Cited above as item C73.

H24 Mehra, Jagdish and Helmut Rechenberg. _The Historical Development of Quantum Theory_. Volume 1. _The Quantum Theory of Planck, Einstein, Bohr,, and Sommerfeld: Its Foundation and the Rise of its Difficulties, 1900-1925_. New York: Springer-Verlag, 1982.

According to the publisher's announcement, this is to be a nine-volume series, covering the quantum theory and its applications to 1965, and the interpretation and epistemology of quantum mechanics from 1926 to present.

H25 Meissner, Walter, et al. "Fifty years of Quantum Theory." _Science_, 113 (1951): 75-101.

Articles by Meissner, Einstein, A. Sommerfeld, F. Bopp, L. Pauling, H. Margenau.

H26 Nicholas, John M. _Anomalies, Falsification, and the History of Science_. Ph. D. Dissertation, University of Pittsburgh, 1976.

Philosophical discussion of aspects of the history of quantum theory.

H27 Nilson, D. R. "Bibliography on the History and Philosophy of Quantum Theory." _Logic and Probability in Quantum Theory_, edited by P. Suppes. Boston: Reidel, 1976, pp. 457-520.

H28 Pais, A. "Einstein and the Quantum Theory." _Reviews of Modern Physics_, 51 (1979): 861-914.

H29 Ponomarev, L. _In Quest of the Quantum_. Moscow: Mir Publishers, 1973. 375 pp.

Translated and revised from the 1971 Russian edition. US distributor: Imported Publications, Inc., 320 West Ohio Street, Chicago, IL 60610.

H30 Ramunni, Girolamo. Les conceptions quantiques de 1911 a 1927. Paris: Vrin, 1981. 196 pp.

"Using published literature (both primary and secondary) the author outlines the historical development, focusing on major philosophical themes around which research centered. The book is intended for a wider audience than merely historians of science."

H31 Rosenfeld, Leon. "Men and ideas in the history of atomic theory." Archive for history of Exact Sciences, 7 (1971): 69-90. Reprinted in item HD49.

On N. Bohr and others.

H32 Sachs, Mendel. The Search for a Theory of Matter. New York: McGraw Hill, 1971. 221 pp.

H33 Sambursky, Shmuel, Ed. Physical Thought from the Presocratics to the Quantum Physicists: An anthology. New York: Pica Press, 1975. xv + 584 pp.

Includes extracts from papers by Planck, Einstein, Rutherford, Bohr, Heisenberg, de Broglie, and Pauli.

H34 Schmutzer, E., ed. 75 Jahre Plancksches Wirkungsquantum,50 Jahre Quantenmechanik. Leipzig: J. A. Barth, 1976. (Nova Acta Leopoldina, Supplementum Nr. 8) 218 pp.

Contains among other essays: F. Hund, "Das Korrespondenzprinzip als Leitfaden zur Quantenmechanik von 1925"; H. Kangro, "Zum Problem der Wäermestrahlungim Vakuum um 1900."

H35 Slater, John C. Concepts and Developments of Quantum Physics. New York: Dover Publications, 1969. xi + 322 pp.

A reprint with "minor corrections" of Slater's Modern Physics (1955).

H36 Szymborski, Krzysztof. Relacje Teorii i Eksperymentu w Genezie Fizyki Kwantowej. Wroclaw / Warszawa / Krakow / Gdansk: Polska Akademia Nauk, Instytut Historii Nauki, Oswiaty i Techniki, Monografie z Dziejow Nauki i Techniki, Tom CXXII (1980). Zaklad Narodowy Imienia Ossolinskich Wydawnictwo Polskiej Akademii Nauk. 198 pp. (Summary in English, pp. 183-85).

H37 Tagliaferri, Guido. *Lineamenti di storia della fisica moderna*. Vol. I: La Nascita della Fisica Quantistica, CLUED, Milano (1977). 237 pp.

Lecture Notes for Physics Major at Milan University.

H38 Tamm, I. E. "Evoliutsii͡a Kvantovoĭ Teorii." *Vestnik AN SSSR*, 9 (1968), 22-28.

On the evolution of quantum theory.

H39 Tarsitani, Carlo. "Le divergenze tra Bohr e Einstein sul significato delle ipotesi quantistiche fino alla formulazione del principio di complementarità." *Contributi* (item H10), pp. 119-218.

H40 Trifonov, D. N. "Background and emergence of the present day atomic theory." *Acta Historiae Rerum Naturalium nec non Technicarum*, 14 (1981): 147-66.

H41 Weislinger, Edmond. *Elements d'histoire et d'epistemologie de la mecanique quantique*. Cahiers d'histoire et de philosophie des sciences, No. 5. Paris: Centre de Documentation Sciences Humaines, 1978. 71 pp.

* Whittaker, Edmund. *A History of the Theories of Aether and Electricity*. Cited above as item F26.

* Wigner, Eugene P. "The unreasonable effectiveness of mathematics in the natural sciences." Cited below as item K108.

H42 Yourgrau, Wolfgang and Stanley Mandelstam. *Variational Principles in Dynamics and Quantum Theory*. Reprint of the third (1968) edition. New York: Dover, 1979. xiii + 201 pp.

Historical review leading up to the use of the principle of least action and similar priniciples in quantum theory.

* * * * * *

For additional references see HW, sections H & L; *Isis Cumulative Bibliography* (item B46), section CC.

HA. BLACK BODY RADIATION AND THE PLANCK-EINSTEIN HYPOTHESES

HA1 Agassi, Joseph. "The Kirchhoff-Planck Radiation Law." Science, 156 (1967): 30-37.

HA2 Alekseev, Igor' Serafimovich. "K predystoriĭ kvantovoi teorii." Voprosy Istorii Estestvoznanii͡a i Tekhniki, 2 (1981): 77-85.

On the pre-history of quantumtheory: black-body radiation, etc.

HA3 Amano, Kiyosi. Netuhukusyaron to Ryosiron no Kigen. 1943.

Book in Japanese on Theory of Heat Radiation and the Origin of Quantum Theory, including Amano's historical introduction discussing the experimental-technological background of Planck's work.

* Bent, Henry A. "Einstein and chemical thought. Atomism extended." Cited above as item DD1.

HA4 Bergia, S., P. Lugli, and N. Zamboni. "Zero-point energy, Planck's law and the prehistory of stochastic electrodynamics. Part I: Einstein and Hopf's paper of 1910." Annales de la Fondation Louis de Broglie, 4, no. 4(1979): 295-318.

HA5 Bergia, S., P. Lugli, N. Zamboni. "Zero-point energy, Planck's law and the prehistory of stochastic electrodynamics. Part 2: Einstein and Stern's paper of 1913." Annales de la Fondation Louis de Broglie, 5, no. 1(1980): 39-62.

HA6 Carazza, Bruno. "Il problema del corpo nero e le origini della meccanica quantistica." Contributi alla storia della meccanica quantistica. Pisa: Domus Galilaeana, Quaderni di Storia e Critica della Scienza, Nuova Serie, 7, 1976, pp. 17-33.

* Compton, Arthur H. "X-rays as a branch of optics." Cited below as item JA4.

HA7 El'i͡ashevich, M. A. "K Istokam Rabot Eĭnshteĭna Po Kvantovoĭ Teorii Izluchenii͡a." Voprosy Istorii Estestvoznanii͡a I Tekhniki, no. 67-68 (1980): 39-45.

The origins of Einstein's work on the quantum theory of radiation.

HA8 Frenk, A. M. "Teoriia Izlucheniia Einshteina." _Einshteinovskii Sbornik, 1971_. Moskva: Nauka, 1972, pp. 192-225.

Einstein's theory of radiation.

HA9 Galison, Peter. "Kuhn and the Quantum Controversy." _British Journal for the Philosophy of Science_, 32 (1981): 71-84.

Discusses item HA22 and the alternative views of M. J. Klein, on Planck.

HA10 Giorello, Giulio. "I concetti di energia, temperatura, entropia in Max Planck (periodo: 1880-1913)." _Physis_, 14 (1972): 211-14.

HA11 Giorello, Giulio. "Le 'ipotesi del disordine' nell'opera di Max Planck: Caos molecolare e radiazione naturale." _Alcuni Aspetti dello Sviluppo delle teorie fisiche 1743-1911_, Saggi di Piero Delsedime et al. Pisa: Domus Galilaeana, 1972, pp. 239-341.

Planck's research in the period 1895-1905.

HA12 Gurikov, V. A. "K voprosy razvitiia teorii teplovogo izlucheniia." _Voprosy Istorii Estestvoznaniia i Tekhniki_, no. 3-4 (56-57)(1977): 69-72.

On the development of heat radiation theory in the late 19th century, including contributions of V. A. Michelson, B. B. Golitsyn, and P. N. Lebedev, leading to Planck's quantum theory.

HA13 Gutting, Gary. "Conceptual structures and scientific change." _Studies in History and Philosophy of Science_, 4 (1973): 209-30.

Uses Planck's quantum theory as an example.

HA14 Hoyer, Ulrich. "Von Boltzmann zu Planck." _Archive for History of exact Sciences_, 23 (1980): 47-86.

HA15 Ivanov, N. I. _Rol' Otechestvennykh Fizikov v Razvitii Kvantovoi Teorii Sveta. Pervaia Chetvert' XX v._ Ulan-Ude, 1969. 255 pp. (Irkut, PI. Buriat. PI. Uchenye Zapiski, Vyp. 32, Fiz.-Mat. Ser., Ch. 1).

The role of Soviet physicists in the development of the quantum theory of light in the first quarter of the20th century.

HA16 Kangro, Hans. _Vorgeschichte der Planckschen Strahlungsgesetzes_. Wiesbaden: Steiner, 1970. xv +

271 pp. Translation: Early History of Planck's Radiation Law. New York: Crane, Russak, 1976. xvii + 282 pp.

HA17 Kangro, Hans. ed. Planck's Original Papers in Quantum Physics. New York: Halsted Press/Wiley, 1972. viii + 60 pp.

The two 1900 papers, annotated, in German and English.

HA18 Kangro, Hans. "Zum Problem der Waermesstrahlung im Vakuum um 1900." (item H34), pp. 21-36.

* Karlov, N. V. and A. M. Prokhorov. "Quantum electronics and Einstein's theory of radiation." Cited below as item HE10.

HA19 Klein, Martin J. "Einstein, Boltzmann's Principle, and the mechanical world view." Proceedings of the XIV International Congress of History of Science, 1974 (pub. 1975), vol. 1, pp. 183-94.

HA20 Klein, Martin J. "The beginnings of the quantum theory." History of Twentieth Century Physics (item A140), pp. 1-39.

HA21 Kobzarev, I. Iu. "A. Einshtein, M. Plank i Atomnaia Teoriia." Priroda, 3 (1979): 8-26.

Einstein, Planck and quantum theory.

HA22 Kuhn, Thomas S. Black-Body Theory and the Quantum Discontinuity. New York: Oxford University Press, 1978. xvi + 356 pp.

Argues that discontinuous energy did not originate with Planck but with Einstein and Ehrenfest. See also the "review symposium" by M. J. Klein, A. Shimony and T. J. Pinch in Isis, 70 (1979): 429-40, and the article by. Gallison (item HA9).

* Millikan, Robert A. "The electron and the light-quant from the experimental point of view." Cited below as item JB18.

HA23 Nickles, Thomas. "Scientific Problems and Constraints." PSA 1978 (Proceedings of the Philosophy of Science Association meeting), vol. 1, pp. 134-48.

On the black body radiation problem, 1859-1900.

HA24 Planck, Max. Die Entdeckung des Wirkungsquantum. Muenchen: Battenberg, 1969. 70 pp.

Reprint of Planck's "Ueber irreversible

Strahlungsvorgaenge", preceded by an essay, "Das Suchen nach dem Absoluten, eine wissenschaftshistorische Einführung," by Armin Hermann.

HA25 Planck, Max. Die Quantenhypothese. Muenchen: Battenberg, 1969. 61 pp.

Reprint of original papers, preceded by an essay, "Die Geburt der Quantentheorie, eine wissenschaftshistorische Einfuehrung" by Armin Hermann.

HA26 Planck, Max. "The genesis and present state of development of the quantum theory." Nobel Lectures...Physics (item A87), Vol. 1, pp. 407-18. Delivered June 2, 1920.

HA27 Rompe, R. "75 Jahre Quantentheorie." Festband zum 75. Jahrestag der Entdeckung der Planckschen Energiequanten. (Abhandlungen,Akademie der Wissenschaften der DDR. Abt. Mathematik, Naturwissenschaften, Technik, N7/1977) Berlin, 1977, pp. 13-24.

HA28 Rosenfeld, L. "Max Planck et la definition statistique de l'entropie." Max-Planck-Festschrift (item 2.P9) 203-11. Translation: "Max Planck and the Statistical Definition of Entropy." Selected Papers of Leon Rosenfeld, edited by R. S. Cohen and J. Stachel. Boston: Reidel, 1979, pp. 235-46.

HA29 Strauss, M. "Max Planck und die Entstehung der Quantentheorie." Forschen und Wirken - Festschrift zur 150-Jahr-Feier der Humboldt-Universität zu Berlin. Bd. I. Berlin, 1960. Translation: "Max Planck and the rise of quantum theory." Modern Physics and its Philosophy. Dordrecht: Reidel, 1972, pp. 23-60.

HA30 Stuewer, Roger H. "Non-Einsteinian interpretations of the photoelectric effect." Historical and Philosophical Perspectives of Science. Edited by Roger H. Stuewer. Minneapolis: University ofMinnesota Press, 1970, pp. 246-63.

The quantum explanation compared to classical interpretations proposed between 1910 and 1913 by H. A. Lorentz, J. J. Thomson, A. Sommerfeld and O. W. Richardson.

HA31 Wien, Wilhelm. "On the laws of thermal radiation." Nobel Lectures...Physics (item A87), Vol. I, pp.275-86. Delivered December 11, 1911.

* * * * * * * * *

For additional references see HW, sections G.q and L.b.

HB. OLD QUANTUM THEORY, ATOMIC MODELS & SPECTROSCOPY BEFORE 1925

HB1 Abraham, Max. "La Dinamica degli elettroni." I progressi recenti (item HB13), pp. 139-161.

HB2 Bellone, Enrico and Nadia Robotti. "Alcuni aspetti dello sviluppo dei primi modelli sulla costituzione dell'atomo." Contributi (item H10), pp. 35-78.

* Belloni, Lanfranco. "Pauli's 1924 note on hyperfine structure." Cited below as item JC8.

HB3 Bohr, Niels. "The structure of the Atom." Nobel Lectures Physics (item 1.N1), vol. 2, 7-43. Delivered December 11, 1922.

HB4 Carazza, Bruno e Guidetti, Gian Paolo. "Ehrenfest e il principio adiabatico." Contributi (item H10), pp. 79-102.

HB5 Casimir, H. B. G. "Walther Nernst und die Quantentheorie der Materie." Berichte der Bunsen-gesellschaft für Physikalische Chemie, 68 (1964): 530-34.

HB6 Cassidy, David C. "Heisenberg's first core model of the atom: The formation of a professional style." Historical Studies in the Physical Sciences, 10 (1979): 187-224.

HB7 Cassidy, David C. "Heisenberg's first paper." Physics Today, 31, no. 7 (July 1978): 23-28.

W. H.'s paper using the "core model" to interpret the anomalous Zeeman effect (1922).

HB8 Darrow, K. K. "The quantum theory." Scientific American, 186, no. 3 (March 1952): 47-54.

History from 1900 to 1923.

HB9 Dingle, Herbert. "A Hundred Years of Spectroscopy." The British Journal for the History of Science, 1 (1963): 199-216.

Includes comments on the work of Alfred Fowler, with whom Dingle worked.

HB10 Enz, Charles P. "Is the zero-point energy real?" Physical Reality and Mathematical Description. Edited by C. P. Enz and J. Mehra. Boston: Reidel, 1974, pp. 124-32.

HB11 Friedman, F. L. and L. Sartori. The Classical Atom. Reading, Mass." Addison-Wesley, 1965. viii + 118 pp.

Historical background of the Rutherford-Bohr model.

HB12 Garbasso, Antonio. "Geometria e cinematica del fenomeno di Zeeman." I progressi recenti (item HB13), pp. 87-106.

HB13 Garbasso, Antonio, ed. I progressi recenti della fisica teorica, sperimentale, applicata. Milano-Roma-Napoli: Dante Alighieri, 1911. 300 pp.

Contains items HB1, HB12, HB40.

HB14 Goudsmit, S. A. "Pauli and nuclear spin." Physics Today, 14, no. 6 (June 1961): 18-21.

Pauli's 1924 paper in Naturwissenschaften introduced nuclear physics into spectroscopy.

HB15 Guidetti, G. P. "Thomson e i modelli atomici." Giornale di Fisica, 20 (1979): 137-142.

Describes Thomson's efforts in building "classical" atomic models.

HB16 Guidetti, G. P. "Un episodio poco noto nella storia della teoria dei quanti: il modello atomico di Whittaker." Giornale di Fisica, 18 (1977): 67-75.

The author discusses E. T. Whittaker's model (1922) and the tentative explanation of line spectra in classical terms.

HB17 Guidetti, G. P. "Anelli di fumo e solitoni." Giornale di Fisica, 17 (1976): 102-108

The author tries to show an analogy between W. Thomson's atomic model (vortex atoms) and the recent idea of describing elementary particles with solitary wave solutions of non linear field-equations.

HB18 Haar, D. ter. "On the history of photon statistics." Quantum Optics. Edited by R. J. Glauber (Rendiconti della Scuola Internazionale di Fisica "E. Fermi," Corso XLII, 1967). New York: Academic Press, 1969, pp. 1-14.

On Planck, Einstein, Ehrenfest and Bose.

HB19 Haas, Arthur Erich. Der Erste Quantenansatz für das Atom. Stuttgart: Battenberg, 1965. 66 pp.

Reprint of his 1910 paper with introduction and bibliography by A. Hermann.

HB20 Heilbron, J. L. "Rutherford-Bohr atom." American Journal of Physics, 49 (1981): 223-31.

HB21 Heilbron, John L. "Lectures on the History of Atomic Physics 1900-1922." _History of Twentieth Century Physics_, (item A140): 40-108.

HB22 Heilbron, John L. _Historical Studies in the Theory of Atomic Structure_. New York: Arno Press, 1981. 302 pp.

The introduction (pp. 1-12) gives a brief account of Pauli's route toward the discovery of the exclusion principle. A collection of reprints of Heilbron's papers follows (on the scattering of alpha & beta particles and Rutherford's atom; genesis of the Bohr atom; the work of H. G. J. Moseley; the Kossel-Sommerfeld theory and the Ring Atom).

HB23 Hendry, John. "Bohr-Kramers-Slater: A virtual theory of virtual oscillators and its role in the history of quantum mechanics." _Centaurus_, 25 (1981): 189-221.

HB24 Hermann, A. "Das Jahr 1913 und der zweite Solvay-Kongress." _Physikalische Blätter_, 19 (1963): 453-462.

HB25 Hermann, A. "Die Entdeckung des Stark-Effektes." _Dokumente der Naturwissenschaft_, 6 (1965): 7-16.

HB26 Hermann, A. _Die Hypothese der Lichtquanten_. Stuttgart: Ernst Battenberg Verlag, 1965. Translation: _Einstein.La teoria dei quanti di luce_. Roma: Newton Compton Italiana, 1973. 84 pp.

HB27 Hermann, A. "Zur Frühgeschichte der Quantentheorie." _Dokumente der Naturwissenschaft_, 7 (1966): 7-22.

HB28 Hermann, A., Ed. _Die Quantentheorie der spezifischen Wärme_. Stuttgart: Ernst Battenberg Verlag, 1967. Translation: _Einstein-Debye-Born-Kármán. La teoria quantistica del calore specifico_. Roma: Newton Compton Editori, 1974. 170 pp.

HB29 Hermann, A. "Die Quantentheorie der spezifischen Wärme." _Dokumente der Naturwissenschaft_, 8 (1967): 7-19.

HB30 Hermann, A. "Die Elektronenstossversuche von Franck und Hertz." _Dokumente der Naturwissenschaft_, 9 (1967): 7-16.

HB31 Hermann, A. "Von Planck zu Bohr. Die ersten fuenfzehn Jahre in der Entwicklung der Quantentheorie." _Angewandte Chemie_, 82 (1970): 1-7. Translation: "From Planck to Bohr. The first fifteen years in the Development of the Quantum Theory." _Angewandte Chemie, International Edition in English_, 9 (1970): 34-40.

HB32 Herzfeld, Karl F. "Bohr atom: A remark on the Early History." Science, 175 (1972): 1393-94.

HB33 Hindmarsh, W. R. Atomic Spectra. Oxford and New York: Pergamon Press, 1967. x + 368 pp.

Includes reprints and translations of papers by Balmer, Rydberg, Bohr, Lorentz, Wentzel, Lande, Sommerfeld and others.

HB34 Hulin, M. "En attendant Debye..." European Journal of Physics, 1 (1980): 222-224.

"The problem of the specific heat of solids and its behaviour at low temperatures played an important role in the evolution of basic ideas in physics at the turn of the century. It was finally solved by Debye, but why did Einstein, who showed a keen interest in that problem for several years fail to propose the Debye model? This article recalls a few facts and presents a few suggestions." (Author's abstract).

HB35 Karlov, N. V., and A. M. Prokhorov. "Kvantovaîa Elektronika i Eĭnshteĭnovskaîa Teoriîa Izlucheniîa." Uspekhi Fizicheskikh Nauk, 128, 3 (1979): 537-543.

Quantum electronics and Einstein's radiation theory.

HB36 Maier, Clifford Lawrence. The Role of Spectroscopy in the Acceptance of an Internally Structured Atom, 1860-1920. Ph.D. Dissertation, University of Wisconsin, 1964. Reprint, New York: Arno Press, 1981. xiv + 601 pp.

HB37 Mehra, J. & H. Rechenberg. Planck's Half-Quanta: History of the Concept of the Zero-Point Energy in the General Development of Quantum Theory. Austin, TX: Center for Particle Theory, 1970.

HB38 Petruccioli, Sandro. "Modello meccanico e regole di corrispondenza nella costruzione della teoria atomica." Physis, 4 (1981): 555-579.

An analysis of the physical and methodological context of the development of N. Bohr's research programme in atomic theory, after 1913.

HB39 Planck, Max et al. "Die ersten zehn Jahre der Theorie von Niels Bohr ueber den Bau der Atome." Die Naturwissenschaften, 11 (1923): 533-624.

Articles by Planck, M. Born, P. Ehrenfest, H. A. Kramers, J. Franck, P. Pringsheim, G. Hertz, D. Coster, A. Kratzer, R. Ladenburg, F. Reiche, W. Kossel, G. V. Hevesy.

HB40 Puccianti, Luigi. "I progressi recenti dell'analisi spettrale." _I progressi recenti_ (item HB13), pp. 107-38.

HB41 Robotti, Nadia. _I Primi Modelli dell' Atomo. Dall'elettrone all'atomo di Bohr_. Torino: Loescher Editore, 1978. 262 pp.

HB42 Robotti, Nadia. "Questioni relative ai primi modelli sulla costituzione dell'atomo." _Seminario di Storia della Fisica..._ (item A110), pp. 53-70.

HB43 Rosenfeld, Leon. "The Wave-Particle Dilemma." _The Physicist's Conception of Nature_ (item A80), pp. 251-63.

HB44 Runge, Iris. "Zur Geschichte der Spektroscopie von Balmer bis Bohr." _Zeitschrift für den Physikalischen und Chemischen Unterricht_, 3 (1939): 103-13.

HB45 Tomonaga, Sin-itiro. _Quantum Mechanics_, Vol. I. _Old Quantum Theory_. Translated from Japanese by Koshiba. Amsterdam: North-Holland Pub. Co., 1962. 313 pp.

HB46 Van der Waerden, B. L. _Sources of Quantum Mechanics_. Amsterdam: North-Holland Pub. Co., 1967. xi + 430 pp.

Includes translations/reprints of original papers by A. Einstein (1916), P. Ehrenfest (1916), N. Bohr (1918, 1924), R. Ladenburg (1921), N. Bohr, H. A. Kramers & J. C. Slater (1924), H. A. Kramers (1924), M. Born (1924), J. H. Van Vleck (1924), H. A. Kramers & W. Heisenberg (1925), W. Kuhn (1925), W.Heisenberg (1925), M. Born & P. Jordan (1925), P. A. M. Dirac (1925, 1926), M. Born, W. Heisenberg & P. Jordan (1925), W. Pauli (1926).

HB47 Zeeman, P. "Les lignes spectrales et les theories modernes de la physique.." _Scientia_, 29 (1921): 13-21.

Spectral analysis from Newton to Rutherford, Bohr and Sommerfeld.

* * * * * * * * * * * * *

For additional references see HW, sections H.a-d, G.e-g, L.c.

HC. QUANTUM MECHANICS AFTER 1920

HC1 Bacry, Henry. "Einstein and de Broglie." Physics Today, 32, no. 8 (August 1979): 11.

Comment on remarks of B. Hoffmann (ibid., no. 3, March 1979, p. 36); with reply by Hoffmann (ibid., August 1979, pp. 11, 13).

HC2 Bellone, Enrico. "Le origini dell'equazione di Dirac - Problemi critici e soluzioni conflittuali, generalizzazione matematica e rielaborazione progressiva dell'evidenze empirica." Contributi...(item H10), pp. 451-65.

HC3 Belloni, Lanfranco. "A note on Fermi's route to Fermi-Dirac statistics." Scientia, 113 (1979): 422-30. Translatiton: "Zametki o puti, privadshem E. Fermi k statistike Fermi-Diraka." Uspekhi Fizicheskikh Nauk, 136 (1982): 167-175.

Fermi's starting point was the 1913 paper of Otto Stern on the chemical constant in the entropy of a gas.

HC4 Besana, Luigi. "La meccanica ondulatoria di Louis de Broglie." Contributi...(item H10), pp. 361-90.

HC5 Bligh, N. M. "The evolution of the new quantum mechanics." Science Progress, 23 (1929): 619-32.

HC6 Born, Max. My Life: Recollections of a Nobel Laureate (item BB15).

Part I, Chapter XIX: "Quantum Mechanics." Part 2, Chapter I: "The 'Heroic Age' of Theoretical Physics."

HC7 Bowen, Marshall, and Joseph Coster. "Born's discovery of the quantum-mechanical matrix calculus." American Journal of Physics, 48 (1980): 491-92.

HC8 (Broglie, Louis de) Louis de Broglie et la mecanique ondulatoire. Choix de textes de Louis de Broglie, edited by M. A. Tonnelat. Paris: Seghers, 1977. 192 pp.

HC9 Broglie, Louis de. Physique et Microphysique. Editions Albin Michel, Paris, 1947. 370 pp.

Chapter VIII contains, "Souvenirs personels sur les debuts de la méchanique ondulaltoire," covering the period 1923-28.

HC10 (Broglie, Louis de) *Vingtieme anniversaire de la mechanique ondulatoire. Plaquette Commemorative publiee sous les auspices du Comite Louis de Broglie*. Paris: Gauthier-Villars, 1944. 119 pp.

Collection of essays by G. Duhamel, A. Lacroix, M. de Broglie and others.

HC11 Broglie, Louis de. "The wave nature of the electron." *Nobel Lectures...Physics*, (item A87), vol. 2, pp. 244-56. December 12, 1929.

HC12 Broglie, Louis de, et al. "K 50 letii͡u stanovlenii͡a kvantovoĭ mekhanike." *Uspekhi Fizicheskikh Nauk*, 122, no. 4 (1977).

Issue commemorating the 50th anniversary of quantum mechanics, with translations of original papers by L. de Broglie, N. Bohr, W. Heisenberg, M. Born, P. Jordan, P. A. M. Dirac and E. Schrodinger.

HC13 Bromberg, Joan. "The concept of particle creation before and after quantum mechanics." *Historical Studies in the Physical Sciences*, 7 (1976): 161-91.

On A. S. Eddington, J. H. Jeans, W. Nernst, P. A. M. Dirac, P. Jordan.

HC14 Caldirola, Piero. "Historical evolution of Exlusion Principle in Physics." *Scientia*, 110 (1975): 69-81.

HC15 Carazza, B. and G. P. Guidetti. "La nascita dell'equazione di Klein-Gordon." *Archive for History of Exact Sciences*, 22 (1980): 373-383.

The authors discuss the different routes to the formulation of the first quantum relativistic wave equation examining the approaches of Klein, Gordon, Fock, de Broglie and Schroedinger.

HC16 Carazza, Bruno and Enrica Giordano. "Schroedinger e le origini della meccanica ondulatoria." *Seminario di Storia della Fisica*...(item A110), pp. 71-95.

HC17 Carazza, Bruno. "Heisenberg - La meccanica delle matrici e il principio di indeterminiazione." *Contributi*.. (item H10), pp. 329-359.

HC18 Casimir, H. B. G. "Some Recollections." *History of Twentieth Century Physics* (item A140), pp. 182-87.

Bohr on electron spin and energy conservation (ca. 1929-30).

HC19 "Cinquantenaire de la mécanique ondulatoire." Revue d'histoire des sciences, 28, no. 1 (1975). 84 pp.

HC20 DeMaria, Michelangelo, Elisabetta Donini, Ester Fano, Paul Forman, John Heilbron, Robert Linhart, Robert Seidel, and Tito Tonietti. Fisica & Societa' Negli Anni '20. Milano: Cooperativa Libraria Universitaria del Politecnico & Cooperativa Libraria Universitaria Editrice Democratica, 1980. 326 pp.

Proceedings of a Workshop at Lecce University on "The Growth of Quantum Mechanics in the 20s and the Cultural, Economic and Social Context of the Weimar Republic and of the U.S.A.," 3-6 September, 1979. For contents in English see Archives Internationales d'Histoire des Sciences, 30 (1980): 181.

Contains item HD23.

HC21 De Maria, M. and F. La Teana. "I primi lavori di E. Schröedinger sulla meccanica ondulatoria e la nascita delle polemiche con la scuola di Göttingen-Copenhagen sull'interpretazione della meccanica quantistica." Physis, 1 (1982): 33-55.

"We analyze the first articles on wave mechanics written by Schroedinger in 1926, in order to single out the main points of divergence - both physical and epistemic - with the Göttingen-Copenhagen (GC) physicists...we stress that the victory of the 'orthodox' interpretation, based on the 'philosophy of observables', implied a deep change in the criteria of scientific explanation." (From authors' abstract).

HC22 Dennison, David M. "Recollections of physics and of physicists during the 1920s." American Journal of Physics, 42 (1974): 1051-56.

On E. Schroedinger, R. Fowler, W. Heisenberg, and his own work on the specific heat of hydrogen.

HC23 Desclaux, J. P. "This Week's Citation Classic: Desclaux J P. Relativistic Dirac-Fock expectation values for atoms with Z=1 to Z=120. Atom.Data Nucl. Data Tables 12: 311-406, 1973." Current Contents, Physical, Chemical and Earth Sciences, 21, no. 34 (Aug. 24, 1981): 16.

The author recalls the circumstances of writing this paper, which has been cited over 140 times since 1973.

HC24 Dirac, Paul A. M. "Theory of electrons and positrons." Nobel Lectures...Physics (item A87), vol. 2, pp. 320-25. December 12, 1933.

HC25 Dirac, P. A. M. "Recollections of an exciting era." _History of Twentieth Century Physics_. (item A140), pp. 109-46.

HC26 Gedei, J. "Polstorocna kvantova mechanika." _Matematika a fyzika ve skole_, 7 (1976/1977): 441-50.

"50 years of quantum mechanics."

HC27 Giordano, Enrica. "Il programma di Werner - Heisenberg tra il 1925 e il 1927." _Seminario di Storia della Fisica_ (item A110), pp. 113-132.

HC28 Goudsmit, Samuel A. "Guess Work: The discovery of the electron spin." _Delta_ (Amsterdam), 15, no. 2 (Summer 1972): 77-91.

Translated from "De Ondekking van de Electronenrotatie," _Nederlands Tijdschrift voor Natuurkunde_, 37, no. 16 (30 September 1971).

HC29 Hankins, Thomas L. "How to get from Hamilton to Schroedinger with the Least Possible Action: Comments on the Optical-Mechanical Analogy." _The Analytic Spirit: Essays in the History of Science_. Edited by Harry Woolf. Ithaca NY: Cornell University Press, 1981, pp. 295-308.

HC30 Heisenberg, Werner. "The development of quantum mechanics." _Nobel Lectures...Physics_ (item A87), vol. 2, pp. 290-301. December 11, 1933.

HC31 Heisenberg, W. "Erinnerungen an die Zeit der Entwicklung der Quantenmechanik." _Theoretical Physics in the Twentieth Century_ (item BP4), pp. 40-47.

HC32 Heisenberg, Werner. "Development of concepts in the History of Quantum Theory." _The Physicist's Conception of Nature_, (item A80), pp. 263-75.

* Hindmarsh, W. R. _Atomic Spectra_. Cited above as item HB33.

Quantum theory of spectral lines, including papers of E. Fermi and E. Segre, G. Racah, G. Breit, H. A. Bethe, V. Weisskopf and E. Wigner.

* Hoddeson, Lillian H. and G. Baym. "The development of the quantum mechanical electron theory of metals: 1900-28." Cited above as item DG37.

HC33 Hoyer, Ulrich. "Wellenmechanik und Boltzmannsche Statistik." _Gesnerus_, 3/4 (1981): 347-349.

Starting from Boltzmann, the author shows a way to find Schroedinger's wave-mechanics.

HC34 Inokuti, Mitio. "This week's citation classic: Inokuti M. Inelastic collisions of fast charged particles with atoms and molecules -- the Bethe theory revisited. Rev. Mod. Phys. 43: 297-347, 1971." Current Contents, Physical Chemical & Earth Sciences, 20, no. 42 (Oct. 20, 1980): 12.

The author recalls the circumstances under which he wrote this article, which has been cited more than 335 times since 1971.

HC35 Jordan, Pascual. "Die Anfangsjahre der Quantenmechanik -- Erinnerungen." Physikalische Blaetter, 31 (1975): 97-103.

See also "Early years of Quantum Mechanics: Some Reminiscences." The Physicist's Conception of Nature (item A80), pp. 294-99.

HC36 Kadomtsev, B. B., V. I. Kogan, B. M. Smirnov, and V. D. Shafranov. "On the Fiftieth anniversary of the paper by M. A. Leontovich and L. I. Mandel'shtam ('On the theory of the Schroedinger equation')." Soviet Physics Uspekhi, 21 (1978): 272-73. Translated from Uspekhi Fizicheskikh Nauk, 124 (1978): 547-57.

The 1928 paper by M. A. L. & L. I. M. treated the transition between discrete and continuous spectra, and "contains within it essentially all the foundations of the theory of penetration of particles through a potential barrier."

* Kevles, Daniel. "The Impact of Quantum Mechanics." The Physicists (item C59), pp. 155-69.

* Konno, Hiroyuki. "Some remarks on the interference experiments with faint light." Cited above as item G20.

HC37 Kronig, R. "The Turning Point." Theoretical Physics in the Twentieth Century (item BP4), pp. 5-39.

Spectroscopic problems in the early 1920s, prehistory of electron spin.

HC38 Kuznetsov, B. G. "Klassicheskie i Neklassicheskie Tendentsii v Volnovoĭ Mekhanike De Broĭla." B. G. Kuznetsov, I. B. Progrebysskiĭ, eds. Frantsuzkaía Nauka i Sovremennaía Fizika. Moskva: Nauka, 1967, pp. 63-79.

Classical and non-classical tendencies in De Broglie's wave mechanics.

HC39 Lestienne, R. and Paty M. "L'edification de la méchanique quantique." *Fundamenta Scientiae*, 17 (1974): 1-37.

HC40 Ludwig, Gunther. *Wave Mechanics*. Oxford & New York: Pergamon Press, 1958. 230 pp.

Anthology of papers by L. de Broglie, E. Schroedinger, W. Heisenberg, M. Born and P. Jordan, with historical introduction.

HC41 Maiocchi, Roberto. "Osservazione sul sorgere della meccanica ondulatoria nelle memorie di Schroedinger del 1926."" *Contributi*...(item H10), pp. 391-406.

HC42 Mehra, Jagdish. *The Birth of Quantum Mechanics*. Geneva: Organisation Europeenne pour la Recherche Nucleaire, 1976 (report CERN 76-10). vi + 56 pp. "Werner Heisenberg Memorial Lecture delivered at the CERN Colloquium on 30 March 1976."

Russian translation in *Uspekhi Fizicheskikh Nauk*, 122 (1977): 719-44.

HC43 Mignani, R. E. Recami and M. Baldo. "About a Dirac-like Equation for the photon, According to Ettore Majorana." *Lettere Nuovo Cimento*, 11 (1974): 568-72.

Following a hint in Ettore Majorana's unpublished notes, the authors establish a Dirac-like equation for the photon.

HC44 Moyer, Donald Franklin. "Origins of Dirac's electron, 1925-1928." *American Journal of Physics*, 49 (1981): 944-49.

HC45 Moyer, Donald Franklin. "Evaluations of Dirac's electron, 1928-1932." *American Journal of Physics*, 49 (1981): 105-62.

HC46 Moyer, Donald Franklin. "Vindications of Dirac's Electron, 1932-1934." *American Journal of Physics*, 49 (1981: 1120-1125.

HC47 Murphy, George. "More on matter waves." *Physics Today*, 27, no. 8 (August 1974): 73-74.

On M. Born and W. Elsasser.

HC48 Niederle, J. "Zrozeni kvantove mechaniky." *Československy časopis pro fyziku*, sekce A. 25 (1975): 392-97.

The birth of quantum mechanics.

HC49 Petruccioli, Sandro. "Il ' principio di corrispondenza': il contributo di Bohr alla fondazione della meccanica delle matrici." Contributi...(item H10), pp. 219-327.

HC50 Polak, L. S. "Vozniknovenie kvantovoi statistiki Boze-Einshteina." Iz Istorii Nauki i Tekhniki v stranakh Vostoka, 1 (1960): 315-29.

The genesis of Bose-Einstein quantum statistics, with a Russian translation of Bose's 1924 paper.

HC51 Price, William C., Seymour S. Chissick and Tom Ravensdale, eds. Wave Mechanics: The First Fifty Years. New York: Wiley/Halsted, 1973. 435 pp.

Includes chapters by L. de Broglie, J. C. Slater, J.H. Van Vleck; most of the articles are on quantum chemistry (includes items HE6, HE20).

HC52 Romer, Reinhold. Der Einfluss Wolfgang Paulis auf die Entstehung der Heisenbergschen Unschäerferelation. Stuttgart: Historische Institut, Zulassungsarbeit, 1977.

HC53 Schroedinger, Erwin. "The fundamental idea of wave mechanics." Nobel Lectures...Physics (item A87), vol. 2, pp. 305-16. December 12, 1933.

HC54 Slater, John C. "Wave Mechanics in the Classical Decade, 1923-1932." Solid-State and Molecular Theory (item BS15), pp. 3-159.

HC55 Tucci, Pasquale. "Il contributo di Dirac alla meccanica quantistica nel periodo '24-'31." Contributi..(item H10), pp. 423-49.

* Van der Waerden, B. L. Sources of Quantum Mechanics. Cited above as item HB46.

HC56 Van der Waerden, B. L. "From Matrix Mechanics and Wave Mechanics toUnified Quantum Mechanics." The Physicist's Conception of Nature (item A80), pp. 276-93.

Includes a letter from W. Pauli to P. Jordan, April 12, 1926.

HC57 Van Vleck, J. H. and D. L. Huber. "Absorption, emission, and linebreadths: A semihistorical perspective." Reviews of Modern Physics, 49 (1977): 939-59.

* Yel'yashevich, M. A. See El'iashevich, M. A.

* * * * * * * * * * * * * * * *

For additional references see HW, sections L.d,e.

HD. PHILOSOPHY OF QUANTUM MECHANICS, HIDDEN VARIABLES

HD1 Alekseev, I. S. Razvitie Predstavleniĭ o Strukture Atoma. Filos Ocherki. Novosibirsk: Nauka, 1968. 123 pp. (AN SSSR, Sibirskoe Otdelenie, Novosibirskiĭ Gosudarstvennyĭ Universitet).

The development of the concept of atomic structure. Philosophical essays.

HD2 Alekseev, I. S. Kontseptsiia Dopolnitel'nosti: Istoriko-Metodologicheskiĭ Analiz. Moskva: Nauka, 1978. 276 pp.

The concept of complementarity: Historical-methodological analysis.

HD3 Alekseev, I. S. "Nekotorye Soobrazheniia po Povodu Diskussii Einshsteina I Bora." Voprosy Filosofii, 1 (1979): 96-103.

That issue carried also papers by V. P. Vizgin, M. E. Omel'ianovskii, B. M. Bolotovskii, V. Ia. Frenkel', D. S. Danin (p.104-120) on different aspects of the Einstein-Bohr discussions.

HD4 Belinfante, F. J. A Survey of Hidden-Variables Theories. New York: Pergamon Press, 1973. 354 pp.

HD5 Bohm, David. "Heisenberg's Contribution to Physics." The Uncertainty Principle and Foundations of Quantum Mechanics. Edited by W. C. Price and S. S. Chissick. New York: Wiley, 1977, pp. 559-63.

HD6 Bohm, David. "Quantum theory as an indication of a new order in physics. Part A. The development of new orders as shown through the history of physics." Foundations of Physics, 1 (1971): 359-81.

* Bohm, David. Causality and chance in Modern Physics. Cited above as item K4.

HD7 Born, Max. "The statistical interpretation of quantum mechanics." Nobel Lectures...Physics (item A87), vol. 3, pp. 256-67. December 11, 1954.

HD8 [Born, Max] Scientific Papers presented to Max Born..on his Retirement from the Tait Chair of Natural Philosophy in the University of Edinburgh. London: Oliver & Boyd, 1953. vi + 94 pp.

Includes articles on the interpretation of quantum mechanics by D. Bohm, L. de Broglie, A. Einstein, A. Lande and E. Schroedinger.

* Born, Max. *Natural Philosophy of Cause and Chance*. Cited below as item K15.

HD9 Brush, Stephen G. "The Chimerical Cat: Philosophy of Quantum Mechanics in Historical Perspective." *Social Studies of Science*, 10 (1980): 394-447.

HD10 Buechel, Wolfgang. "Der Bellsche Beweis. Eine Fallstudie." *Zeitschrift fuer allgemeine Wissenschaftstheorie*, 8 (1977): 221-36.

Contrary to Kuhn & Lakatos, a "purely metaphysical" thesis became falsifiable and falsified by a "crucial" experiment, so that followers of an older paradigm changed their opinion on rational grounds.

HD11 Bunge, Mario. "Survey of the interpretations of Quantum Mechanics." *American Journal of Physics*, 24 (1956): 272-86.

HD12 Caldirola, Piero. "Old and new Problems in the Interpretation of Quantum Theory." *Scientia*, 110 (1975): 701-735.

HD13 Cassirer, Ernst. *Determinismus und Indeterminismus in der Modernen Physik. Historische und Systematische Studien zum Kausalproblem* (Goeteborgs Högskolas Årsskrift XLII, 1936: 3). Goeteborg: Elanders Boktryckeri Aktiebolag, 1937. 265 pp. Translation: *Determinism and Indeterminism in Modern Physics: Historical and Systematic Studies of the Problem of Causality*. Translated by O. Theodore Benfey, with a preface by Henry Margenau. New Haven: Yale University Press, 1956. xxiv + 227 pp.

HD14 Costa de Beauregard, O. "Le paradoxe d'Einstein, Podolsky et Rosen." *Bulletin de la Société Francaise de Philosophie*, 71, no. 1 (1977): 1-30.

HD15 Costa de Beauregard, Olivier. "The 1927 Einstein and 1935 EPR Paradox." *Physis*, 22 (1980): 211-242.

"The so-called Einstein paradox is thus shown to be born from the union of two earlier paradoxes: 1) *intrinsic time symmetry*, as discussed by Loschmidt and Zermelo in classical statistical mechanics; 2) *Born's principle of adding partial amplitudes* rather than probabilities, forbidding to think of micro-entities as objects separately *endowed* with *properties*. The [R.P.] Feynman zigzag thus appears as the *deus-ex-machina* of the Einstein correlation...

HD16 DeWitt, B. S. and R. N. Graham. "Resource Letter IQM-1 on the Interpretation of Quantum Mechanics." *American Journal of Physics*, 39 (1971): 724-38.

Annotated bibliography of selected works on the subject.

HD17 Dirac, P. A. M. "The evolution of the physicist's picture of Nature." *Scientific American*, 208, no. 5 (May 1963): 45-53.

HD18 Eddington, A. S. "The decline of determinism." *Mathematical Gazette*, 16 (1932): 66-80.

HD19 Faggiani, D. "Fisica quantistica e tradizione filosofica." *Scientia*, 106 (1971): 993-1003. English translation: "Quantum physics and the philosophical tradition." *Scientia*, 106 (1971): 1004-13.

HD20 Feyerabend, Paul K. *Realism, Rationalism and Scientific Method, Philosophical Papers*, Vol. 1. New York: Cambridge University Press, 1981. xiv + 353 pp.

Includes papers on the interpretation of quantum mechanics by N. Bohr and H. Reichenbach.

HD21 Folse, Henry J. "Platonic 'atomism' and contemporary physics." *Tulane Studies in Philosophy*, 27 (1978): 69-88.

Based on a comparison of Democritean and Platonic views with the modern theory of elementary particles, the author concludes that the latter is closer in spirit to the Platonic view since it claims to give "a *myth* -- a likely story in exactly the same sense as does the *Timaeus* -- and not an *ontology* of 'what is' as Democritean atomism intends to give; but it is a 'myth' which, unlike Plato's...is not *merely* speculative but is a speculation anchored to experimental corroboration." (p. 75)

HD22 Forman, Paul. "Weimar culture, causality, and quantum theory, 1918-1927: Adaptation by German physicists and mathematicians to a hostile intellectual environment." *Historical Studies in thte Physical Sciences*, 3 (1971): 1-115.

A classic paper frequently cited by sociologists of science as an illustration of the influence of culture on scientific ideas.

HD23 Forman, Paul. "Kausalitaet, Anschaulichkeit und Individualitaet oder Wie Wesen und Thesen, die der Quantenmechanik Zugeschrieben, durch Kulturelle Werte Vorgeschrieben Wurden." Koelner Zeitschrift fuer Soziologie und Sozialpsychologie. Sonderheft 22: Wissenssoziologie. Edited by Nico Stehr and Volker Meja. Wiesbaden: Westdeutscher V., 1980, pp. 393-406. Translation: "Kausalitaet, Anschaulichkeit e Individualitaet; ovvero, come i valori culturali prescrissero il carattere e gli insegnamenti attribuiti alla meccanica quantistica." Fisica & Societa' negli Anni '20 (item HC20), pp. 15-34.

Slightly revised and enlarged version of Forman's talk at the Lecce 1979 "Workshop on the Growth of Quantum Mechanics in the Cultural, Economic and Social Context of the Weimar Republic.

HD24 Geymonat, Ludovico. Storia del pensiero filosofico e scientifico. Milano: Garzanti, 1970-1976.7 volumes, 5248 pp.

HD25 Giordano, Enrica. "Schroedinger e l'interpretazione fisica della meccanica quantistica." Contributi...(item H10), pp. 407-421.

HD26 Graham, Loren R. Science and Philosophy in the Soviet Union. New York: Knopf, 1972. xii + 584 pp.

Chapter III, "Quantum Mechanics, 69-110.

* Harvey, Bill. "The effects of social context on the process of scientific investigation: experimental tests of quantum mechanics." Cited below as item JE12.

* Harvey, Bill. "Plausibility and the evaluation of knowledge: A Case-study of experimental quantum mechanics." Cited below as item JE13.

HD27 Heelan, Patrick A. "Heisenberg and radical theoretical change." Zeitschrift fuer allgemeine Wissenschaftstheorie, 6 (1975): 113-136.

Claims that Heisenberg's research was guided by certain methodological principles. With comments by Heisenberg on his alleged disagreements with Bohr (ibid. p. 137)

HD28 Heisenberg, Werner. "Bemerkungen ueber die Entstehung der Unbestimmtheitsrelation." Physikalische Blaetter, 31 (1975): 193-96.

HD29 Hendry, John. "Weimar culture and quantum causality." *History of Science*, 18 (1980): 155-80.

Critique of Forman's theses (see item HD22).

HD30 Hoerz, Herbert. *Werner Heisenberg und die Philosophie*. Berlin: VEB Deutscher Verlag der Wissenschaften, 1968. 308 .pp

HD31 Holton, Gerald. "The Roots of Complementarity." *Daedalus*, 99 (1970): 1015-55. Reprinted in his *Thematic Origins of Scientific Thought* (item A46)

Relates the origin of Bohr's complementarity principle to his readings of philosophical works by P. M. Moller, William James, H. Hoffding, Kierkegaard.

HD32 Hooker, C. A., Ed. *The Logico-Algebraic Approach to Quantum Mechanics*. Volume I. *Historical Evolution*. Boston: Reidel, 1975. xi + 607 pp.

Reprints of papers by Birkhoff and von Neumann (1936), Strauss (1937), Reichenbach (1944), Putnam (1957), Feyerabend (1958) and others on quantum logic.

HD33 Hooker, C. A. "The nature of Quantum Mechanical Reality: Einstein versus Bohr." *Paradigms and Paradoxes*. Edited by R. G. Colodony. Pitsburgh: University of Pittsburgh Press, 1972, pp. 67-302.

HD34 Huebner, Kurt. "The philosophical Background of Hidden Variables in Quantum Mechanics." *Man and World*, 6 (1973): 421-40.

HD35 Jammer, Max. "Indeterminacy in physics." *Dictionary of the History of Ideas*. Edited by P. Weiner. New York: Scribner, 1973, Vol. II, pp. 585-94.

HD36 Jammer, Max. *The Philosophy of Quantum Mechanics: The Interpretations of Quantum Mechanics in Historical Perspective*. New York: Wiley, 1974. 536 pp.

Comprehensive treatment, fully documented.

HD37 Kedrov, B. M. and N. F. Ovchinnikov, eds. *Printsip Sootvetstvii͡a: Istoriko-Metodologicheskiĭ Analiz*. Moskva: Nauka, 1979. 317 pp.

Principle of correspondence: historico-methodological analysis.

HD38 Krajewski, Wladislaw. *Correspondence Principle and the Growth of Science*. Boston: Reidel,1977. xiv + 136 pp.

HD39 Levy, Edwin, Jr. *Interpretations of Quantum Theory and Soviet Thought*. Dissertation, Indiana University, 1969.

For summary see *Dissertation Abstracts International*, 30 (1960): 5482-A.

HD40 Mackay, Donald M. "Religion and the new mechanics." *Janus*, 64 (1977): 119-29.

Theological implications of indeterminism and complementarity.

HD41 Mehra, Jagdish. "The Quantum Principle: Its Interpretation and Epistemology." *Dialectica* 27 (1973): 75-157.

Views of E. Wigner. Also a separate monograph with the same title, Boston: Reidel Publishing Company, 1974.

HD42 Omelyanovsky, M. E. *Dialectics in Modern Physics*. Moscow: Progress Pubs., 1979.

On the philosophy of quantum mechanics.

HD43 Pinch, Trevor J. "What does a proof do if it does not prove? A Study of the Social Conditions and Metaphysical Divisions leading to David Bohm and John von Neumann failing to Communicate in Quantum Physics." *The Social Production of Scientific Knowledge*. Edited by E. Mendelsohn et al. Boston: Reidel, 1977, pp. 171-215.

HD44 Primas, Hans. *Chemistry, Quantum Mechanics, and Reductionism*. Berlin & New York: Springer-Verlag, 1981. xii + 451 pp.

Semi-historical/philosophical treatment of quantum mechanics.

HD45 Reichenbach, Hans. *Philosophische Grundlagen der Quantenmechanik*. Basel: Verlag Birkhaeuser,1949. 198 pp. Translation: *I fondamenti filosofici della meccanica quantistica*, Torino: 1964. 307 pp.

HD46 Roeseberg, Ulrich. *Determinismus und Physik*. Berlin: Akademie-Verlag, 1975. 208 pp.

Philosophical analysiss of epistemological problems of quantum mechanics in a "diamat" perspective.

HD47 Roeseberg, Ulrich. *Quantenmechanik und Philosophie*. Berlin: Akademie-Verlag, 1978. 226 pp.

Discusses philosophical background of birth and development of non-relativistic quantum mechanics from a "diamat" perspective.

HD48 Rosen, Nathan. "Can Quantum-Mechanical Description of Physical Reality be considered Complete?" Albert Einstein: His Influence on Physics, Philosophy and Politics (item BE9), pp. 57-67.

A review of the paper with this title published by Albert Einstein, B. Podolsky and N. Rosen in 1935, discussing N. Bohr's reply to it.

HD49 [Rosenfeld, Leon] Selected Papers of Leon Rosenfeld. Edited by R. S. Cohen and J. J. Stachel. Boston: Reidel Publishing Co., 1969. xxxiv + 929 pp.

Includes: "The Epistemological Conflict between Einstein and Bohr" (1963); "Niels Bohr's Contribution to Epistemology" (1963).

HD50 Scheibe, Erhard. "Bibliographie zu Grundlagen-fragen der Quantenmechanik." Philosophia Naturalis, 10 (1968): 249-50.

A list of about 600 items on philosophy and logical foundations of quantum mechanics.

HD51 Stent, Gunther S. "Does God Play Dice?" The Sciences, 19, No. 3 (March 1979): 18-23.

The real issue in the Bohr-Einstein debate was monotheism (E.) vs. atheism (B.)

HD52 Stolzenburg, Klaus. Die Entwicklung des Bohrschen Komplementaritaetsgedankens in den Jahren 1924 bis 1929. Stuttgart: Historisches Institut, Dissertation, 1975.

HD53 Suvorov, S. G. "Problema 'Fizicheskoĭ Real'-nosti' v Kopengagenskoĭ Shkole." Uspekhi Fizicheskikh Nauk. 62 (1957): 141-58.

On the works of M. Born on philosophical problems of physics in the previous 30 years.

HD54 Tagliagambe, Silvano. "L' interpretazaion della meccanica quantistica in URSS alla luce del materialismo dialettico." L. Geymonat et al. , Storia del pensiero filosofico e scientifico, Vol. VI (Il Novecento), pp. 729-61.

HD55 Teller, Paul. "The projection postulate and Bohr's interpretation of quantum mechanics." PSA 1980 (in press).

* Toulmin, Stephen, ed. Physical Reality. Cited above as item K102.

HD56 Unruh, W. G. and G. I. Opat. " The Bohr-Einstein 'weighing of energy' debate." American Journal of Physics, 47 (1979): 743-44.

HD57 Vucinich, Alexander. "Soviet Physicists and Philosophers in the 1930s: Dynamics of a Conflict." Isis, 71 (1980): 236-50.

Mostly on the Copenhagen Interpretation.

HD58 Wallace, William A. Causality and Scientific Explanation. Volume 2, Classical and Contemporary Science. Ann Arbor: University of Michigan Press, 1974. xi + 422 pp.

HD59 Walsh, Francis Augustine. "The new physics and scholasticism." Aspects of the New Scholastic Philosophy. Edited by Charles A. Hart. New York: Benziger Bros., 1932, pp. 40-59.

A discussion of philosophical aspects of atomic physics in relation to scholasticism.

HD60 Weinzierl, Emil. "Zur sowjetphilosophischen Kritik des quantenmechanischen Indeterminismus." Zeitschrift fur Philosophische Forschung, 32 (1978): 109-23.

* * * * *

For additional references see HW, sections L.i and W.c.

HE. APPLICATIONS OF QUANTUM THEORY TO OTHER SCIENCES

HE1 Abe, Yuko. "Pauling's revolutionary role in the development of quantum chemistry." _Historia Scientiarum_, 20 (1981): 107-24.

HE2 Alekseev, G. N. "Energeticheskie epokhi i osnovanye periody razvitiiâ îadernoĭ energetiki." _Voprosy Istorii Estestvoznaniiâ i Tekhniki_, 2 (1981), 121-129.

"Power epochs and basic periods of development of nuclear power engineering."

HE3 Ballhausen, C. J. "Quantum mechanics and chemical bonding in inorganic complexes. 1: static concepts of bonding; dynamic concepts of valency. 2: Valency and inorganic metal complexes. 3: The spread of the ideas." _Journal of Chemical Education_, 56 (1979): 215-18, 294-97, 357-61.

HE4 Bantz, David A. "The Structure of Discovery: Evolution of Structural Accounts of Chemical Bonding." _Scientific Discovery: Case Studies_, edited by Thomas Nickles. Dordrecht & Boston: Reidel Publishing Company, 1980, pp. 291-329.

Assessment of the contributions of Werner Heitler, Fritz London, and G. N. Lewis.

HE5 Chandrasekhar, S. "The case for astronomy." _Proceedings of the American Philosophical Society_, 108 (1964): 1-6.

On R. H. Fowler's application of Fermi-Dirac statistics to the theory of white dwarf stars (1926).

HE6 Coulson, Charles A. "The influence of wave mechanics on organic chemistry." _Wave Mechanics: The First Fifty Years_ (item HC51), pp. 255-71.

HE7 Dmitriev, I. S. and S. G. Semenov. _Kvantovaia Khimia - Eë Proshloe i Nastoîashchee_. Moskva: Atomizdat, 1980. 160 pp.

Historical discussion of valency, quantum numbers, molecular orbitals, chemical bonds, etc. from the 19th century to the present.

HE8 (Fukui, K.) Streitweser, Andrew, Jr. "The 1981 Nobel Prize in Chemistry." _Science_, 214 (1981): 627-29.

On the award to Kenichi Fukui and Roald Hoffmann for their quantum mechanical studies of chemical reactivity.

HE9 Graham, Loren R. "A Soviet Marxist view of structural chemistry: The theory of resonance controversy." Isis, 55 (1964): 20-31.

* [Hoffmann, R.] Streitweser, Andrew, Jr. "The 1981 Nobel Prize in Chemistry." Cited above as item HE8.

HE10 Karlov, N. V., and A. M. Prokhorov. "Quantum electronics and Einstein's theory of radiation." Soviet Physics Uspekhi, 22 (1979): 576-79. Translated from Uspekhi Fizicheskikh Nauk, 128 (1979): 537-43.

HE11 Kaufman, John. "Criticism of the theory of resonance in organic chemistry 1944-1956." Synthesis, 4, no. 2 (1977): 44-49.

HE12 Kimball, G. E. "Reminiscences of 1933-1935." Quantum Chemistry, 1 (1967): 845-47.

HE13 Leicester, Henry M. Source Book in Chemistry 1900-1950. Cambridge, Mass.: Harvard University Press, 1968. xvii + 408 pp.

Includes extracts of papers on atomic and molecular structure.

HE14 Mulliken, Robert S. "Spectroscopy, molecular orbitals, and chemical bonding." Nobel Lectures...Chemistry (item A86), vol. 4 (Dec. 12, 1966), pp. 131-60.

HE15 Nachtrieb, Norman H. "Interview with Robert S. Mulliken." Journal of Chemical Education, 52 (1975): 560-64.

On Mulliken's career, application of quantum mechanics to chemistry.

HE16 Paradowski, Robert J. The Structural Chemistry of Linus Pauling. Ph.D. Dissertation, University of Wisconsin, 1973. 579 pp.

For summary see Dissertation Abstracts International, 33 (1973): 3026B.

* Price, W. C. et al., eds. Wave Mechanics, The First Fifty Years. Cited above as item HC51.

HE17 Rich, A. & N. Davidson, eds. Structural Chemistry and Modern Biology: A Volume Dedicated to Linus Pauling by his students, colleagues, and friends. San Francisco: Freeman, 1968. 907 pp.

Includes articles on Pauling's work in quantum chemistry.

HE18 Rompe, R. "Zur Geschichte der Beziehungen der Quantenphysik und der Technik in den ersten Jahrzehnten des 20.Jh." _75 Jahre Quantentheorie_. (Abhandlungen der Akademie der Wissenschaften der DDR, Abt. Mathematik, Naturwissenschaften, und Technik, N. 7) Berlin, 1977, pp. 13-24.

HE19 Stent, G. S. "That was the molecular biology that was." _Science_, 160 (1968): 390-95.

On the influence of N. Bohr, E. Schroedinger and M. Delbruck.

HE20 Urch, David S. "Influence of Wave Mechanics on Inorganic Chemistry." _Wave Mechanics: The First Fifty Years_. (item HC51), pp. 160-73.

* * * * * *

For additional references see HW, section R.b

HF. FIELD THEORY AND QUANTUM ELECTRODYNAMICS

HF1 Anderson, Davis K. et al. "More tests of QED." Physics Today, 26, no. 2 (February 1973): 13, 15.

HF2 Belloni, L. "Historical remarks on the 'classical' electron radius." Lettere al Nuovo Cimento, 31 (1981): 131-34.

HF3 Cassidy, David C. "Cosmic ray showers, high energy physics, and quantum field theories: Programmatic interactions in the 1930s." Historical Studies in the Physical Sciences, 12 (1981): 1-39.

HF4 Peierls, R. E. "Field theory since Maxwell." Clerk Maxwell and Modern Science. Edited by C. Domb. London: University of London, Athlone Press, 1963, pp. 26-42.

HF5 Schwinger, Julian, Ed. Selected Papers on Quantum Electrodynamics. New York: Dover Publications, 1958. xvii + 424 pp.

Reprints of papers by P. A. M. Dirac (1927, 1933, 1934), E. Fermi(1930), Dirac et al.(1932), P. Jordan & E. Wigner (1928), W. Heisenberg (1934), V. F. Weisskopf (1936, 1939), J. R. Oppenheimer (1950), S. Tomonaga (1946, 1948), J. Schwinger (1948, 1949, 1951,1953), W. Pauli & F. Villars (1949), R. P. Feynman (1948, 1949,1950), F. J.Dyson (1949), W. Pauli (1940), R. Karplus & A. Klein (1952), G. Kallen (1953), N. M. Kroll & W. E. Lamb, Jr. (1949).

HF6 Schwinger, Julian. "A Report on Quantum Electrodynamics." The Physicist's Conception of Nature (item A80), pp. 413-29.

HF7 Tomonaga, Sin-itiro. "Development of Quantum Electrodynamics." Physics Today, 19 , no. 7 (Sept. 1966): 25-32. Reprinted in The Physicist's Conception of Nature (item A80), pp. 404-12.

HF8 Weisskopf, Victor F. "The development of field theory in the last 50 years." Physics Today, 34, no. 11 (Nov. 1981): 69-85.

HF9 Wentzel, Gregor. "Quantum theory of fields (until 1949)." Theoretical Physics in the Twentieth Century (item BP4), pp. 48-77. Reprinted in item A80.

* * * * * *

For additional references see HW, Section L.h

JA. X-RAYS

JA1 Barkla, Charles G. "Characteristic Roentgen radiation." _Nobel Lectures...Physics_ (item A87), Vol. I, pp. 393-99. Delivered June 3, 1920.

JA2 Bragg, William Lawrence. "The diffraction of X-rays by crystals." _Nobel Lectures...Physics_. (item A87), Vol. I, 370-82. Delivered September 6, 1922.

JA3 Bragg, W. L. "The diffraction of Röntgen rays by crystals." _Beiträge zur Physik und Chemie des 20. Jahrhunderts_ (item B11), pp. 147-51.

JA4 Compton, Arthur H. "X-rays as a branch of optics. _Nobel Lectures...Physics_ (item A87), vol. 2, pp. 174-90. Delivered December 12, 1927.

JA5 Feather, Norman. "Historical Introduction to Alembic Club Reprint, no. 22." _X-rays and the Electronic Conductivity of Gases_. Edinburgh: Livingstone, 1958, pp. 1-27.

JA6 Forman, Paul. The discovery of the diffraction of X-rays by crystals: A critique of the Myths." _Archive for History of Exact Sciences_, 6 (1969): 38-71.

Followed by P. P. Ewald's comments, ibid. 72-81. See also item JA7.

JA7 Gasman, L. D. "Myths and X-rays." _British Journal of the Philosophy of Science_, 26 (1975): 51-60.

Critique of the interpretation of Forman (item JA6).

JA8 Gerlach, Walther. "Die Roentgenstrahlen in der Geschichte der Wissenschaft." _Strahlentherapie_, 140 (1970): 245-51.

JA9 Glusker, Jenny P., Ed. _Structural Crystallography in Chemistry and Biology_. (Benchmark Papers in Physical Chemistry and Chemical Physics, Volume 4) Stroudsburg, PA: Hutchinson Ross, 1981. 448 pp.

Includes M. von Laue, "Historical Introduction," and reprints of papers by W. Friedrich, P. Knipping and M. Laue, P. P. Ewald, W. L. Bragg, W. H. Bragg, L. Pauling,. and others.

JA10 Hildebrandt, G. "62 Jahre Kristalloptik der Roentgenstrahlen I. Ewalds Theorie." Physikalische Blätter, 35, no. 2 (Feb. 1979): 55-64.

Review of Ewald's dynamic theory of 1917 describing interference in an idealized perfect crystal lattice and extending optical interference theory to X-ray wavelengths.

JA11 Hildebrandt, G. "62 Jahre Kristalloptik der Röntgenstrahlen II. Die Folgen." Physikalische Blätter, Vol. 35, no. 3 (March 1979): 103-118.

Deals with confirmation and extension of the basic theory due to Ewald from 1920-40. Deals also with investigations of diffraction in an ideal crystal from 1940-1955 and the subsequent development of X-ray topography (A. R. Lang).

* Kirkpatrick, Paul. "Confirming the Planck-Einstein equation $h - (1/2)mv^2$." Cited below as item JD9.

JA12 Laue, Max von. "Concerning the detection of X-ray interferences." Nobel Lectures...Physics (item A87), Vol. I, pp. 347-55. Delivered November 12, 1915.

JA13 Maurer, Hans-Joachim and Werner Weber. "Die Entdeckung der Rontgenstrahlen Trivialliteratur und der Fachpresse von 1896-1901." Technikgeschichte, 44 (1977): 324-39.

JA14 Muldawer, Leonard. "Resource letter XR-1 on X rays." American Journal of Physics, 37 (1969): 123-34.

Annotated bibliography of sources.

JA15 Pavlova, G. E. "Iz Istorii Organizatsii Sovetskoĭ Rentgenologii. 1919-1925 gg." Materialy Godichnoĭ Konferentsii Leningradskogo Otdeleniîa Sovetskogo Natsional'nogo Ob'edineniîa Isotrikov Estestvoznaniîa i Tekhniki. Leningrad, 1968, pp. 24-25.

History of organization of Soviet Roentgenology, 1919-1925.

JA16 Roentgen, Wilhelm Conrad. Uber eine neue Art von Strahlen. Mit einem biographischen Essay von Walther Gerlach. Hrsg. und mit einem Vorwort versehen von Fritz Krafft. Munchen: Kindler, 1972. 106 pp.

JA17 (Roentgen, W. C.) Debye, Peter. "Roentgen und seine Entdeckung" Deutsche Museum, Abhandlungen und Berichte, 1934. 21 pp.

* Siegbahn, Manne. "The X-ray spectra and the structure of the atoms." Cited below as item JD15.

JA18 Wheaton, Bruce R. "Impulse x-rays and radiant intensity: the double edge of analogy." Historical Studies in the Physical Sciences, 11 (1981): 367-90.

On J. J. Thomson, W. Wien, A. Sommerfeld, M. v. Laue.

JA19 Wiederkehr, Karl Heinrich. "Uber die Entdeckung der Roentgenstrahl-interferenzen durch Laue und die Bestaetigung der Kristallgittertheorie." Gesnerus, 3/4 (1981): 351-69.

Treats "the efforts to explain the nature of X-rays, their character of e.m. waves and at the same time the lattice structure of crystals and related controversies between atomists and dynamists."

JA20 Wyckoff, R. W. G. "Some thoughts on the future of natural science, with illustrations from the growth of X-ray diffraction work in the United States." British Journal of Applied Physics, 5 (1954): 199-204.

JA21 Zaǐtseva, L. L. "Pervaia Radiologicheskaia Laboratorii͡a v. Rossii." Trudy Instituta Istorii Estestvoznanii͡a i Tekniki AN SSSR, Moskva, 19 (1957): 197-218.

On the first Radiological Lab established in Odessa in 1910.

* * * * *

For additional references see HW, sections B.rk7 and G, k-o; Isis CB, "Personalities" under Roentgen and "Subjects," section CX.

JB. THE ELECTRON (DISCOVERY, CHARGE, ETC.)

JB1 Anderson, David L. "Resource Letter ECAN-L on the Electronic Charge and Avogadro's Number." American Journal of Physics, 34 (1966): 2-8.

Annotated bibliography.

JB2 Anderson, David L. "Cathode Rays and the Discovery of the Electron." Discoveries in Physics (item JF5), Chapter 2.

JB3 Arons, A. B. "Phenomenology and logical reasoning in introductory physics courses." American Journal of Physics, 50 (1982): 13-20.

Includes a discussion of J. J. Thomson's experiment establishing the particle nature of the electron.

* Bader, Morris. "The Nobel Prize of 1973." Cited below as item JD1.

JB4 Bailey, Paul T. "Discovery of the electron." Physics Today, 19, no. 7 (July 1966): 12.

Early measurements of e/m.

JB5 Battimelli, G. "The electromagnetic mass of the Electron: A case study of a non-crucial experiment." Fundamenta Scientiae, 2 (1981): 137-50.

On the experiments of W. Kaufmann (1906) and their failure to settle the choice between the theories of M. Abraham (1902-3), Lorentz (1904), and Einstein (1905).

* Belloni, L. "Historical remarks on the 'classical' electron radius." Cited above as item HF2.

JB6 Bruzzaniti, Giuseppe. "L'introduzione del cronone nella teoria classica dell'elettrone. Analisi di alcuni aspetti storici ed epistemologici." Epistemologia, 4 (1981): 381-406.

Philosophical discussion of the "chronon" hypothesis of Caldirola, "which originated historically as a development of the Lorentzian programme aimed at establishing the equation of motion of an accelerated electron."

* Davisson, Clinton J. "The discovery of electron waves." Cited below as item JE8.

JB7 Dukov, V. M. Elektron. Istoriia Otkrytiia i Izucheniia Svoistv. Moskva: Prosveshchenie, 1966. 235 pp.

History of the electron's discovery and of the study of its properties.

JB8 Fairbank, William M., Jr. and Allan Franklin. "Did Millikan observe fractional charges on oil drops?" American Journal of Physics, 50 (1982): 394-97.

"We have reanalyzed Millikan's 1913 data on oil drops to examine the evidence in favor of charge quantization and no convincing evidence for fractional residual charge on the oil drops." (Authors' summary)

JB9 Fletcher, Harvey. "My work with Millikan on the Oil-Drop Experiment." Physics Today, 35, no. 6 (June 1982): 43-47.

"In this personal reminiscence the late author recounts his experiencesas a graduate student in the Ryerson laboratory in Chicago and his contribution to the determination of the electron's charge. Harvey Fletcher (1884-1981) directed acoustical and, later, physical research at Bell Laboratories from 1925 to 1952, developing hearing aids and stereophonic equipment." (Note by editor of Physics Today; the manuscript of this paper was provided by Mark B. Gardner, a long-time friend of Fletcher.)

JB10 Franklin, Allan. "Millikan's published and unpublished data on oil drops." Historical Studies in the Physical Sciences, 11 (1981): 185-201.

"Although Millikan claimed that the 58 drops published in his famous 1913 paper were his entire set of data, this is not, in fact the case. All of Millikan's oil drop data in the period Oct. 28, 1911 until April 16, 1912, have been reanalyzed to examine whether he selectively analyzed his data. I conclude that he was selective in both his choice of and his analysis of his data, but that the effects were, in general, quite small." (Author's summary)

JB11 Holton, Gerald. "Subelectrons, presuppositions, and the Millikan-Ehrenhaft dispute." Historical Studies in the Physical Sciences, 9 (1977): 161-224. Reprinted in The Scientific Imagination (item A48), pp. 25-83.

JB12 Holton, Gerald. "Electrons or subelectrons? Millikan, Ehrenhaft and the role of preconceptions." History of Twentieth Century Physics (item A140), pp. 266-89.

JB13 Kleen, Werner. "Horst Rothe, 1899-1974." Physikalische Blaetter, 31 (1975): 39-40.

JB14 Kusch, Polykarp. "The magnetic moment of the electron." Nobel Lectures..Physics (item A87), vol. 3, pp. 298-310. December 12, 1955.

JB15 Lenard, Philipp E. A. "On cathode Rays." Nobel Lectures...Physics (item A87), Vol. I, pp. 105-34. Delivered May 28, 1906.

JB16 Lorentz, H. A. "The theory of electrons and the propagation of light." Nobel Lectures...Physics (item A87), Vol. I, pp. 14-29. Delivered December 11, 1902.

JB17 Malley, Marjorie. "The discovery of the beta particle." American Journal of Physics, 39 (1971): 1454-60.

Russian translation in Uspekhi Fizicheskikh Nauk, 109 (1973): 389-98.

JB18 Marton, L., and C. Marton. "Evolution of the concept of the elementary charge." Advances in Electronics and Electron Physics, 50 (1980): 449-72.

JB19 Millikan, Robert A. "The electron and the light-quanta from the experimental point of view." Nobel Lectures...Physics (item A87), Vol. 2, pp. 54-66. Delivered May 23, 1924.

JB20 Morrow, B. A. "On the discovery of the electron." Journal of Chemical Education, 46 (1969): 584-88.

On the significance of the experiments of J. J. Thomson and R. A. Millikan.

JB21 Pais, A. "The early history of the theory of the electron: 1897-1947." Aspects of Quantum Theory (item BD6).

JB22 Pyenson, Lewis. "Physics in the shadow of mathematics: the Goettingen electron-theory seminar of 1905." Archive for History of Exact Sciences, 21 (1979): 55-89.

JB23 Rohrlich, Fritz. "The Electron: Development of the First Elementary Particle Theory." The Physicist's Conception of Nature (item A80), pp. 331-69.

JB24 Romer, Alfred. "The experimental history of atomic charge, 1895-1903." Isis, 34 (1942): 150-61.

JB25 Schaffner, Kenneth F. "The Lorentz electron theory of relativity." American Journal of Physics, 37 (1969): 498-513.

According to the author, the title should have been "...and relativity."

JB26 Spivak, G. V. "Gazovaia elektronika." Razvitie Fizike v Rossii (item A93), pp. 99-117.

JB27 Ternov, I. M. and V. A. Borodvitsyn. "Modern interpretation of J. I. Frenkel's classical spin theory." Soviet Physics Uspekhi, 23 (1980): 679-83. Translated from Uspekhi Fizicheskikh Nauk, 132 (1980): 345-52.

On generalizations of Frenkel's 1926 equation.

* Thomson, George P. "Electronic waves." Cited below as item JE25.

JB28 Thomson, Joseph J. "Carriers of negative electricity." Nobel Lectures...Physics (item A87), Vol. I, pp. 145-53. Delivered December 11, 1906.

JB29 Vyaltsev, A. N. "The discovery of electron and its scientific and technical implications." Acta Historiae Rerum Naturalium nec non Technicarum, 14 (1981): 227-48.

JB30 Zahar, Elie. "'Crucial' experiments: A case study." Progress and Rationality in Science. Edited by G. Radnitzky and G. Andersson. Boston Studies in the Philosophy of Science, 58. Boston/Dordrecht: Reidel, 1978, pp. 71-97.

On the 1905 experiment by W. Kaufmann, intended to decide between the theories of the electron proposed by Abraham and by Einstein & Lorentz.

* * * * * * * * * * * * *

For additional references see HW, section E.b.

JC. RADIOACTIVITY, COSMIC RAYS, PARTICLE AND NUCLEAR PHYSICS TO 1930

JC1 Amaldi, E. "Radioactivity, a pragmatic pillar of probabilistic conceptions." Problems in the Foundations of Physics. Proceedings of the International School of Physics "Enrico Fermi," Course XLLII. Edited by G. Toraldo di Francia. New York: North-Holland Publ. Co., 1979, pp. 1-28.

The author traces the origins of probabilistic conceptions to early discoveries in radioactivity.

JC2 Aston, Francis William. "Mass spectra and isotopes." Nobel Lectures...Chemistry (item A86), Vol. 2, pp. 7-20. December 12, 1922.

JC3 Badash, Lawrence. Radioactivity in America: Growth and Decay of a Science. Baltimore: Johns Hopkins University Press, 1979. xi + 327 pp.

Detailed account of the research of B. B. Boltwood, H. A. Bumstead, H. N. McCoy, E. Rutherford, F. Soddy; includes concurrent European research.

JC4 Badash, Lawrence. "The suicidal success of radiochemistry." British Journal for the History of Science, 12 (1979): 245-56.

On the work of K. Fajans, F. Soddy; isotopes & atomic weights.

JC5 Badash, Lawrence. "Radium, Radioactivity, and the Popularity of Scientific Discovery." Proceedings of the American Philosophical Society, 122 (1978): 145-54.

JC6 Batalin, A. Kh. "Razvitie Radiokhimicheskogo Analiza v SSR. 1917-1946." Mariia Sklodovskaia-Kiuri. 1867-19657. Materialy i Tezisy Konferentsii. Orenburg, 1968. 15-20.

On the development of radiochemical analysis in USSR from 1917 to 1946.

JC7 Becquerel, Antoine H. "On radioactivity, a new property of matter." Nobel Lectures...Physics (item A87), Vol. 1, pp. 52-70. December 11, 1903.

JC8 Belloni,, Lanfranco. "Pauli's 1924 note on hyperfine structure." American Journal of Physics, 50 (1982): 461-464.

"It has seemed paradoxical that in 1924, some four months before he dismissed Kronig's tentative suggestion of a spinning electron, Pauli had proposed in print that a nonvanishing angular momentum be ascribed to the nucleus. The purpose of the present note is to resolve the paradox..."

JC9 Bruzzaniti, Giuseppe. La Radioattivita' da Becquerel a Rutherford. Torino: Loescher, 1980. 22 pp.

Anthology of excerpts from Roentgen, Poincare, P. and M. Curie, Debierne, Rutherford, Soddy and Bachelard on nuclear transmutations.

JC10 Bykova-Orlova, E. G. "Pervye Issledovaniia po Radioaktinvosti v Rossii." Tezisy Dokladov i Soobshchenii na Meshvuzovskoi Konferentsi po Istorii Fiziko-Matematicheskikh Nauk. Moskva, 1960. 113 pp.

On the beginning of radioactivity research in Russia, early 20th century.

JC11 Condon, Edward U. "Tunneling--how it all started." American Journal of Physics, 46 (1978): 319-23.

Text of a talk at University of Colorado, March 8, 1969, on the contributions of Condon and Ronald Gurney to the theory of nuclear decay by quantum tunneling.

JC12 Cranston, J. A. The discovery of isotopes by Soddy and his school in Glasgow. (Address to the Royal Philosophical Society) Glasgow, 1954. 12 pp.

JC13 Curie, Marie Sklodowska. "Radium and the new concepts in chemistry." Nobel Lectures...Chemistry (item A86), Vol. 1, pp. 202-12. December 11, 1911.

JC14 Curie, Pierre. "Radioactive substances, especially radium." Nobel Lectures...Physics (item A87), Vol. I, pp. 73-78. Delivered June 6, 1905.

JC15 Darwin, C. G. "The discocvery of Atomic Number." Niels Bohr and the Development of Physics. Edited by W. Pauli. New York: McGraw-Hill, 1955, pp. 1-11.

Brief review of the contributions of F. Soddy, J. J. Thomson, E. Rutherford, N. Bohr, H. Moseley.

JC16 Dezso, Erwin. Ket emberper négy tudós három Nobeldij Pierre Curie, Scolodovska, Irene Curie, Frédéric Joliot-Curie. Valogatta, forditottá, bevezétessel és utószóval ellátta Deszo Erwin. Bukarest: Kriterion Konyvkiadó, 1970. 168 pp.

JC17 Dorman, I. V. Kosmicheskie Luchi. (Istoricheskii Ocherk) Nauka: Moskva, 1981. 191 pp.

Historical essay on cosmic ray research, very competently written especially on Soviet contributions.

JC18 Feather, N. "Some episodes of the alpha-particle story, 1903-1977." Proceedings of the Royal Society A357 (1977): 117-29. Reprinted in Rutherford and Physics at the Turn of the Century (item BR12), pp. 74-88.

Personal recollections of E. Rutherford, his experiments, and his reactions to quantum theory.

JC19 Ginzburg, Vitalii Lazarevich and Irina Vitalievna Dorman. "Prioroda i Proiskhozhdenie Kosmicheskikh Lucheĭ: Istoriia i Sovremennost'." Priroda, 4 (1978), 10-29.

Nature and origin of cosmic rays: past and present.

JC20 Goldansky, V. I. and D. N. Trifonov. "Sem'desiat piat' let ucheniya o radioaktivnosti." Voprosy Istorii Estestvoznaniia i Tekhniki, 1 (34) (1971): 3-12.

"75 Years of Radioactivity" with a chronology of events, ibid., pp. 13-19.

JC21 Grigorov, N. L. and L. G. Mishchenko. "Kosmicheskie Luchi." Razvitie Fiziki v Rossii (item A93), pp. 256-83.

JC22 Grinberg, A. P. "History of the invention and development of accelerators (1922-1932)." Soviet Physics Uspekhi, 18 (1975): 8815-31. Translated from Uspekhi Fizicheskikh Nauk, 117 (1975): 333-62.

Survey of early work on accelerators by Soviet physicists L. V. Mysovskii, G. I. Babat, K. D. Sinel'nikov, A. K. Val'ter, and Western physicists Ising, Breit, Tuve, Slepian, Lange, Wideroe, Rutherford, Coolidge, Szilard, etc.

JC23 Hirosige, Tetu. "The Van den Broek hypothesis." Japanese Studies in the History of Science, 10 (1971): 143-62.

Proposal that electric charge of nucleus is equal to atomic number. A short versison of this paper appeared in Proceedings of the XIII International Congress of the History of Science, Moscow, 1971, vol. 6, pp. 97-101.

JC24 Iakovlev, K. P. "K Istorii Pervykh Rabot po Radioaktivnosti v Fizicheskom Institute MGU (1900-1930gg.)." Istoriia i Metodologiia Estestvennykh Nauk, 2 (1963): 298-307.

Includes a bibliography of Russian works on radioactivity at Moscow State University from 1900 to 1930.

JC25 Iakovlev, K. P. "Raboty po Radioactivnosti v Fizicheskom Institute MGU. (Period 1900-1930gg.)." Tezisy Dokladov i Soobshchenii Na Mezhvuzovskoi Konferentsii po Istorii Fiziko-Matematicheskikh Nauk. Moskva, 1960. 132 pp.

Radioactivity researches at Moscow State University from 1900 to 1930.

JC26 Jenkins, E. N. Radioactivity: A science in Historical and Social Context. New York: Crane, Russak, 1979. viii + 197 pp. (Second edition of book published in 1964 as An Introduction to Radioactivity.)

JC27 Kargon, Robert. "Birth cries of the elements: Theory and experiment along Millikan's route to cosmic rays." The Analytic Spirit: Essays in the History of Science in Honor of Henry Guerlac. Edited by Harry Woolf. Ithaca, NY: Cornel University Press, 1981, pp. 309-25.

JC28 Kedrov, B. M., ed. Uchenie o Radioaktivnosti. Istoriia i Sovremennost'. Moskva: Nauka, 1973. 254 pp.

Investigations on radioactivity. Historical and contemporary.

JC29 Keller, Cornelius. Die Geschichte der Radioaktivitaet. Unter besonderer Beruecksichtigung der Transurane. Stuttgart: Wissenschaftliche Verlagsgesellschaft, 1982. 112 pp.

JC30 Krivomazov, A. N. "The importance of the discovery of radioactivity and radium in the context of the scientific-technological progress." Acta Historiae Rerum Naturalium nec non Technicarum, 14 (1981): 167-86.

JC31 Malley, Marjorie. "The discovery of atomic transmutation: Scientfic Styles and Philosophies in France and Britain." Isis, 70 (1979): 213-23.

On the failure of P. Curie and A. Debierne to arrive at the transmutation theory.

JC32 Ostroumov, B.A. V. I. Lenin i Nizhegorodskaia Radiolaboratoriia Istoriia Laboratorii v Dokumentakh i Materialakh. Leningrad: Nauka,1967. 407 pp. (AN SSSR, Institut Istorii Estestvoznaniia i Tekhniki).

Documents pertaining to V. I. Lenin's action in support of Nizhegorod Radiolaboratory.

JC33 Pais, A. "Radioactivity's two early puzzles." Transactions of the New York Academy of Sciences, 38 (1977): 116-36. Reviews of Modern Physics, 49 (1977): 925-38.

Energy source; significance of half-life.

JC34 Richards, Theodore William. "Atomic weights." Nobel Lectures...Chemistry (item A86), Vol. 1, pp. 280-92. December 6, 1919.

JC35 Romer, Alfred, Ed. The Discovery of Radioactivity and Transmutation. New York: Dover, 1964. xi + 233 pp.

Reprints of translations, with commentary, of papers by H. Becquerel (1896, 1901), E. Rutherford (1900, 1905), W. Crookes (1899-1900), E. Rutherford and F. Soddy (1902, 1903), P. & M. Curie (1902), P. Curie & A. Laborde (1903), W. Ramsay & F. Soddy (1903).

JC36 Romer, Alfred, Ed. Radiochemistry and the Discovery of Isotopes, With commentary and an Introductory Historical Essay. New York: Dover Pubs., 1970. xiii + 261 pp.

Includes papers by Marie Curie and Pierre Curie (1898-1906), F. Fiesel (1899), W. Marckwald (1902-1910), O. Hahn (1905-8), B. B. Boltwood (1907-1908), H. N. McCoy & W. H. Ross (1907), F. Soddy (1911-13), K. Fajans (1913), A. Fleck (1913).

JC37 Rutherford, Ernest. "The chemical nature of the alpha particles from radioactive substances." Nobel Lectures...Chemistry (item A86), Vol. 1, pp. 129-40. December 11, 1908.

JC38 Rutherford, Ernest. "Forty Years of Physics." Background to Modern Science (item A85), pp. 49-74.

JC39 Rutherford, Ernest; Bertram B. Boltwood. Rutherford and Boltwood: Letters on radioactivity. Edited by Lawrence Badash. New Haven: Yale University Press, 1969. xxii + 378 pp. (Yale studies in the history of science and medicine)

Includes all known letters between Ernest Rutherford and Bertram B. Boltwood, covering the period from 1904 to 1924.

* Segre, Emilio. <u>From X-Rays to Quarks</u>. Cited above as item A107.

JC40 Soddy, Frederick. "The origins of the conceptions of isotopes." <u>Nobel Lectures..Chemistry</u> (item A86) Vol. 1, pp. 371-99. December 12, 1922.

JC41 Spronsen, Jan W. Van. "Atomic number before Moseley." <u>Journal of Chemical Education</u>, 56 (1979): 106.

On A. J. van den Broek.

JC42 Squire, William. "H. G. Wells foresees isotopes." <u>Physics Today</u>, 30, no. 8 (August 1977): 66.

JC43 Starosel'skaia-Nikitina, O. A. and E. A. Starosel'skaia. "Rol'Marii Sklodovskoi-Kiuri v Razvitii Iadernoi Fiziki." <u>Priroda</u>, 2 (1968): 90-96.

On the role of Maria Sklodowska-Curie in the development of nuclear physics.

JC44 Trenn, Thaddeus J. <u>Transmutation: Natural and Artificial</u>. Philadelphia: Heyden & Son, Inc. 1981. 160 pp.

A history of transmutation with emphasis on three Nobel Prize-winning papers by Rutherford & Soddy (1908), Curie & Joliot (1934), and Hahn & Strassman (1938).

JC45 Trenn, Thaddeus J. "Thoruranium (U-236) as the extinct natural parent of thorium: the premature falsification of an essentially correct theory." <u>Annals of Science</u>, 34 (1978): 581-97.

On the suggestion of Gerhard Kirsch (1922), rejected but revived after 1940.

JC46 Trenn, Thaddeus J. <u>The Self-Splitting Atom: The History of the Rutherford-Soddy Collaboration</u>. London: Taylor & Francis, 1977. 175 pp.

JC47 Trifonov, D. N., and G. A. Khakimbaeva. "Ob otkrytii iskusstvennogo prevrashcheniia elementov." <u>Voprosy Istorii Estestvoznaniia i Tekhniki</u> 60, No. 3 (1978): 80-84.

"On the discovery of the artificial transmutation of elements." Concludes that April, 1919 is the date only for the discovery of artificial nuclear reactions (without any detailed study of their mechanism); the real proof for artificial transmutation was only obtained in 1925.

JC48 Vernov, S. N. and N. A. Dobrotin. "Fiftieth anniversary of a fundamental discovery in cosmic-ray physics." Soviet Physics Uspekhi, 20 (1977): 531-35.

On the 50th anniversary of D. V. Skobel'tsyn's paper (1927) on cosmic ray physics (study of charged particles in cosmic rays).

JC49 Vial'tsev, A. N., A. N. Krivomazov, and D. N. Trifonov. Pravilo sdviga i iavlenie isotopii. Moscow: Atomizdat, 1976. 208 pp.

The displacement rule and the phenomena of isotopes.

JC50 Wilson, Charles T. R. "On the cloud method of making visible ions and the tracks of ionizing particles." Nobel Lectures...Physics (item A87), vol. 2, pp. 194-214. Delivered December 12, 1927.

JC51 Wright, Stephen, ed. Classical scientific papers--Physics. New York: American Elsevier, 1965. xix + 393 pp.

Facsimile reproductions of papers by Rutherford and Soddy (1902), Rutherford (1903, 1906, 1914, 1920), Rutherford & T. Royds (1909), J. J. Thomson (1897, 1898, 1899), Rutherford and J. M. Nuttall (1913), H. Geiger & E. Marsden (1913), Moseley (1913, 1914), J. Chadwick (1932), O. W. Richardson & K. T. Compton (1912), J. D. Cockcroft & E. T. S. Walton (1932), A. H. Compton & R. L. Doan (1925), C. T. R. Wilson (1897, 1912), Aston (1933).

JC52 Zaitseva, L. L., Osnovnye Etapy Razvitiia Ucheniia O Radioaktivnosti v Dorevoliutsionnoi Rossii. Moskva, Izd-Vo AN SSSR, 1957. 25 pp.

Basic landmarks of development of radioactivity studies in pre-revolutionary Russia.

* * * * *

For additional references see HW, sections I and K.a, c; Isis CB, sections CS, CU.

JD. ATOMIC PHYSICS EXPERIMENTS BEFORE 1926

JD1 Bader, Morris. "The Nobel Prize of 1923." _Journal of Chemical Education_, 55 (1978): 783.

Work of Wilson, Przibram, Millikan and Ehrenhaft on electron charge.

JD2 Bothe, Walther, "The coincidence method.." _Nobel Lectures...Physics_ (item A87), pp. 271-76.

Prize for 1954, lecture not given orally owing to his illness.

* Chwolson, O. D. _Die Physik 1914-1926_. Cited above as item A23.

JD3 Compton, A. H. "The scattering of X rays as particles." _American Journal of Physics_, 29 (1961): 817-20.

Recalls his discovery of the "Compton effect" and controversy with Duane in 1923-24.

JD4 Flecken, F. A. "Gaede's influence on the development of mechanical vacuum pumps." _Vacuum_, 13 (1963): 583-88.

JD5 Franck, James. "Transformations of kinetic energy of free electrons into excitation energy of atoms by impacts." _Nobel Lectures...Physics_ (item A87), vol. 2, pp. 98-108. December 11, 1926.

JD6 Hertz, Gustav. "The results of the electron-impact tests in the light of Bohr's theory of atoms." _Nobel lectures...Physics_ (item A87), vol. 2, pp. 112-29. December 11, 1926.

JD7 Kachele, Volker. _Zeeman- und Starkeffekt als Prufstein der Bohrschen Atomtheorie_. Stuttgart: Historisches Institut, Zulassungsarbeit, 1970.

JD8 Kerker, Milton. "The Svedberg and molecular reality." _Isis_, 67 (1976): 190-216.

JD9 Kirkpatrick, Paul. "Confirming the Planck-Einstein equation $h\nu = (1/2)mv^2$." _American Journal of Physics_ 48 (1980): 803-6.

In a 1964 interview, David L. Webster II stated that he had done the 1915 experiment (discovery of inverse photoelectric effect and first x-ray determination of _h_) for which W. Duane and F. L. Hunt received the credit.

JD10 Kovacs, I. "The role of atomic spectra in the study of the structure of the atom." (In Hungarian), _Fizikai Szemie_, vol. 23, no. 3, (March 1973): 65-71.

JD11 Maier, Clifford L. _The role of spectroscopy in the acceptance of the internally structured atom, 1860-1920_. New York: Arno Press, 1981.

Reprint of his Ph.D. Dissertation at University of Wisconsin, Madison, 1964.

JD12 Matsnev, A. M. "K Istorii Izucheniia Gazovogo Razriada v Rossii v XIX i Nachale XX v." _Tezisy Dokladov i Soobshchenii na Mezhvuzovskoi Konferentsii po Istorii Fiziko-Matematicheskikh Nauk_. Moskva: Nauka, 1960, pp. 107-108.

On the history of gas discharge studies in Russia, 19th and early 20th centuries.

JD13 Raevskii, I. P. "Razvitie Spektral'noi Tekhniki v Pervom Desiatiletii XX Veka." _Voprosy Istorii Fizikiiee Prepodavaniia_ (1961): 131-136.

Development of spectral techniques in the period 1900-1910.

JD14 Rosmorduc, Jean. "Une erreur scientifique au debut de siecle: 'Les Rayon N.'" _Revue d'Histoire des Sciences et de leurs Applications_, 25 (1972): 13-25.

On R. Blondlot.

JD15 Siegbahn, Manne. "The X-ray spectra and the structure of the atoms." _Nobel Lectures...Physics_ (item A87), Vol. 2, pp. 81-89. December 11, 1925.

JD16 Stark, Johannes. "Structural and spectral changes of chemical atoms." _Nobel Lectures...Physics_ (item A87), Vol. I, pp. 427-35. June 3, 1920.

JD17 Stern, Otto. "The method of Molecular rays." _Nobel Lectures...Physics_ (item A87), vol. 3, pp. 8-16. (Prize for 1943, lecture delivered December 12, 1946).

JD18 Stuewer, Roger H. _The Compton Effect: Turning Point in Physics_. New York: Science History Publications, 1975. xii + 367 pp.

Gives a comprehensive account of developments in radiation physics leading up to and influenced by the discovery by A. H. Compton in 1922; shows that the discovery was not directly motivated by a desire to test Einstein's quantum hypothesis. The author also discusses the work of C. G. Barkla, N. Bohr, W. H. Bragg, P. Debye, E. Rutherford, A. Sommerfeld, J. J. Thomson and D. L. Webster.

JD19 Thuillier, P. "La triste histoire des rayons N." _La Recherche_, 95 (1978): 1092-1101.

Discusses the mistaken discovery of R. Blondlot in 1903.

JD20 Trigg, George L. _Crucial Experiments in Modern Physics_. New York: Van Nostrand Reinhold, 1971. ix + 141 pp.

Black-body radiation, transmutation of elements, existence of atoms (Perrin), atomic nucleus (Geiger and Marsden), collisions of electrons with atoms (Franck and Hertz), photoelectric effect (Millikan), spatial orientation of atomic magnets (Stern and Gerlach), particle properties of light (A. H. Compton).

* Wilson, Charles T. R. "On the cloud method of making visible ions and the tracks of ionizing particles." Cited above as item JC50.

* Wright, Stephen, Ed. _Classical scientific papers--Physics_. Cited above as item JC51.

JE. ATOMIC PHYSICS EXPERIMENTS AFTER 1925

JE1 Anderson, Davis K., Richard T. Robiscoe and John M. Wessner. "More tests of QED." *Physics Today*, 26, no. 2 (February 1973): 13, 15.

JE2 Bederson, Benjamin. "Atomic physics: a renewed vitality." *Physics Today*, 34, no. 11 (Nov. 1981): 188-209.

Survey of developments in the U.S. since about 1950.

JE3 Berlaga, R. Ia.; V. N. Vertsner; A. A. Lebedev. "Elektronnaiâ Mikroskopiiâ v Sovetskom Soiuze." *Zavodskaiâ Laboratoriiâ*, 10 (1957): 1214-1219.

Electron microscopy in USSR.

JE4 (Bloembergen, N.) Stoicheff, Boris P. "Laser spectroscopy shares prize." *Science*, 214 (1981): 631-33.

On the award of half of the 1981 Nobel Prize to Nicolaas Bloembergen and Arthur L. Schawlow.

* Bothe, Walther. "The coincidence method." Cited above as item JD2.

JE5 Brix, Peter. "50 Jahre Kernvolumeneffekt in den Atomspektren." *Physikalische Blaetter*, 37 (1981): 181-3.

On the 1931 discovery of the isotopic displacements in spectra of heavy elements by H. Schuler, J. E. Keyston and H. Kopfermann.

JE6 Cosslett, V. E. "The development of electron microscopy and related techniques at the Cavendish Laboratory, 1947-79." *Contemporary Physics*, 22 (1981): 3-36, 147-82.

JE7 Cosslett, V. E. "Electron microscopy: Retrospect and prospect. A personal history of the subject." *Proceedings of the Royal Microscopic Society*, 14 (1979): 101-7.

On his work at the Cavendish Laboratory, 1946-78.

JE8 Davisson, Clinton J. "The discovery of electron waves." *Nobel Lectures...Physics* (item A87), vol. 2,pp. 387-94. December 13, 1937.

JE9 Eckertova, L. "Vedecka prace elektroniky a vakuove fyziky matematicko-fyzikalni fakulty UK Praha. (K 25. vyroci MFF UK)." Pokroky matematiky, fyziky a astronomie, 23 (1978): 111-13.

"Scientific research in electronics and vacuum physics at the Faculty of Mathematics and Physics of Charles University, Prague. (25th anniversary of faculty)"

JE10 Gehrenbeck, Richard K. "Electron diffraction: Fifty Years Ago." Physics Today, 31, no. 1 (January 1978): 34-41.

See also comment by R. Schlegel, ibid., no. 7 (July 1978): 9, 11, and reply by Gehrenbeck, ibid. 11, 13, on M. Born and W. Elsasser.

JE11 Goodman, P., Ed. Fifty Years of Electron Diffraction. Boston: Reidel, 1981. xiv + 440 pp.

Includes: H. A. Medicus, "The Origin of de Broglie's concept," 7-11; R. K. Gehrenbeck, "Davisson and Germer," 12-27; P. B. Moon, "George Paget Thomson," 28-39; S. Miyake, "Seishi Kikuchi," 40-54; short papers by L. de Broglie, M. J. H. Ponte, C. J. Calbick, H. A. Bethe, J. J. Trillat, M. Blackman, H. Mark, L. O. Brockway, L. E. Sutton, S. H. Bauer, G. Haegg, S. Miyake, R. Uyeda, Y. Morino, T. Hibi, S. Ogawa, Z. G. Pinsker, H. Wilman, H. Boersch, P. Goodman, C. H. MacGillavry, E. G. McRae, G. A. Somorjai, V. Schomaker, K. Hadberg, O. Bastiansen, L. S. Bartell, J. Karle, I. Karle, G. Honjo, H. Hashimoto, K. Tanaka, J. M. Cowley, S. Moss, J. V. Sanders, P. B. Hirsch, L. Sturkey, F. Fujimoto, B. B. Zvyagin, S. A. Semiletov, R. M. Imamov, L. V. Vilkov.

JE12 Harvey, Bill. "The effect of social context on the process of scientific investigation: Experimental Tests of Quantum Mechanics." The Yearbook of the Sociology of the Sciences, 5 (1981): 139-63.

On the experiment of R. A. Holt and its reception by physicists.

JE13 Harvey, Bill. "Plausibility and the evaluation of knowledge: A Case-study of Experimental Quantum Mechanics." Social Studies of Science, 11 (1981): 95-130.

On the "hidden variables" experiments done in the 1970s, and how they were interpreted by physicists (work of J. S. Bell, S. J. Freedman and J. F. Clauser, R. A. Holt and F. Pipkin).

* Inokuti, Mitio. "This Week's Citation Classic: Inokuti M. Inelastic collisions of fast charged particles with atoms and olecules -- the Bethe theory revisited...." Cited above as item HC34.

JE14 Kopferman, Hans, et al. Physics of the Electron Shells. (FIAT Review of German Science, 1939-1946). Wiesbaden: Office of Military Government for Germany, Field Information Agencies Technical, British, French, U. S., 1948. 130 pp. (Text in German).

* Kusch, Polykarp. "The magnetic moment of the electron." Cited above as item JB13.

JE15 Lafferty, James M. "Vacuum: From art to exact science." Physics Today, 34, no. 11 (Nov. 1981): 211-31.

Historical survey of vacuum technology for the special issue "50 years of physics in America."

JE16 Lamb, Willis E., Jr. "Fine structure of the hydrogen atom." Nobel Lectures...Physics (item A87), vol. 3, pp. 286-95. December 12, 1955.

JE17 Lamb, W. E., Jr. "Some history of the hydrogen fine structure experiment." A Festschrift for I. I. Rabi (item BR2), pp. 82-86.

* Lubkin, G. B. "Nobel Physics Prize to Bloembergen, Schawlow and Siegbahn." Cited below as item G29.

JE18 Moessbauer, Rudolf L. "Recoilless nuclear resonance absorption of gamma radiation." Nobel Lectures...Physics (item A87), vol. 3, pp. 584-601. December 11, 1961.

* Motz, Lloyd, Ed. A Festschrift for I. I. Rabi. Cited above as item BR2.

JE19 Mulvey, T. "Forty years of electron microscopy." Physics Bulletin, 24 (1973): 147-54.

JE20 Raman, Chandrasekhara V. "The molecular scattering of light." Nobel Lectures...Physics (Item A87), vol. 2, pp. 267-75. December 11, 1930.

JE21 Rohrlich, Fritz. "Delbruck scattering." Physics Today, 34, no. 12 (Dec. 1981): 72-73.

Disagrees with G. Stent (ibid., June, p. 71) who said Bethe demonstrated the existence of the phenomenon.

JE22 Russo, Arturo. "Fundamental research at Bell Laboratories: The discovery of electron diffraction." _Historical Studies in the Physical Sciences_, 12 (1981): 117-60

On the work of C. J. Davisson and L. H. Germer.

* (Schawlow, A. L.) Stoicheff, Boris P. "Laser spectroscopy shares prize." Cited above as item JE4.

JE23 Series, G. W. "Thirty years of optical pumping." _Contemporary Physics_, 22 (1981): 487-509.

Survey of the work of A. Kastler (1950), J. Brossel & F. Bitter (1952) on optical pumping, optical "magnetization" of gases and vapours, spin exchange, metastability exchange, orientation by collision, multiple quantum resonances, atomic coherences, quantum beats, dressed atoms and laser spectroscopy.

JE24 [Siegbahn, K. M.] Hollander, Jack M., and David A. Shirley. "The 1981 Nobel Prize in Physics." _Science_ 214(1981): 629-31.

On the award of half of the prize to Kai M. Siegbahn for his work on electron spectroscopy.

JE25 Thomson, George P. "Electronic waves." _Nobel Lectures...Physics_ (item A87), vol. 2, pp. 397-403. Delivered June 7, 1938.

* Trigg, George L. _Crucial Experiments in Modern Physics_ (Cited above as item JD20).

Chapter 10, "Wave properties of matter."

JE26 Vertsner, V. N. "Raboty GOI v Oblasti Elektronnoi Mikroskopii." _Trudy Gosudarstvennogo Opticheskogo Instituta_, 27, no. 156 (1960): 24-53.

Research on electron microscopy at Moscow State Optical Institute from 1939-1960.

JF. NUCLEAR AND PARTICLE PHYSICS 1930-1945; ATOMIC BOMB

JF1 Akchurin, I. A. "Razvitie Fiziki Antichastits." Voprosy Filosofii, 6 (1957): 159-163.

On the development of anti-particle physics since the late '20s.

JF2 Amaldi, Edoardo. "Venticinque anni dalla prima reazione a catena divergente controllata." Cultura e scuola, 26 (1968): 5-13.

Celebration of 25th anniversary of first controlled chain reaction, with recollections of Fermi's work.

JF3 Amaldi, Edoardo. "Personal Notes on Neutron Work in Rome in the 30s and Post-war European Collaboration in High-Energy Physics." History of Twentieth Century Physics (item A140), pp. 294-351.

JF4 Anderson, Carl D. "The production and properties of positrons. Nobel Lectures...Physics (item A87), vol. 2 pp. 365-76. December 12, 1936.

JF5 Anderson, David L. Discoveries in Physics. (The Project Physics Course, Supplemental Unit B) New York: Holt, Rinehart and Winston, 1973. iii + 92 pp.

Chapter 3, "Nuclear Fission"; Chapter 4, "The Neutrino."

JF6 Blackett, Patrick M. S. "Cloud chamber researches in nuclear physics and cosmic radiation." Nobel Lectures...Physics (item A87), vol. 3, pp.97-119. December 13, 1948.

JF7 Bothe, Walther, and Siegfried Fluegge. Nuclear Physics and Cosmic Rays. (FIAT Review of German Science, 1939-1946). (Wiesbaden:) Office of Military Government for Germany, Field Information Agencies Technical, British, French, U.S., 1948. 2 vols.

JF8 Brown, Laurie M. "Yukawa's prediction of the meson." Centaurus, 25 (1981): 71-132.

JF9 Brown, L. M., M. Konuma and Z. Maki, Eds. Particle Physics in Japan, 1930-1950. Kyoto: Research Institute for Fundamental Physics, 1980. 2 vols. iv + 71; iv + 71 pp.

JF10 Brown, Laurie M. and Lillian Hoddeson. " The birth of elementary-particle physics." Physics Today, 35, no. 4 (April 1982): 36-43.

* Cassidy, David C. "Cosmic ray showers, high energy physics, and quantum field theories: Programmatic interactions in the 1930s." Cited above as item HF3.

JF11 Chadwick, James. "The neutron and its properties." *Nobel Lectures...Physics* (item A87), vol. 2, pp. 339-48. Delivered December 12, 1935.

JF12 Clark, R. W. *The Birth of the Bomb: The untold story of Britain's part in the weapon that changed the world.* London: Phoenix House, 1961. 209 pp. New York: Horizon Press, 1961 (American edition does not have subtitle).

JF13 Clark, Ronald W. *The Greatest Power on Earth. The international Race for Nuclear Energy*. New York: Harper & Row, 1980. ix + 342 pp.

JF14 Cockcroft, John D. "Experiments on the interaction of high-speed nucleons with atomic nuclei." *Nobel Lectures...Physics* (item A87), vol. 3, pp. 167-84. December 11, 1951.

JF15 Conversi, Marcello, Ed. *Evolution of Particle Physics: A Volume dedicated to Edoardo Amaldi in his 60th Birthday*. New York: Academic Press, 1970. xxii + 342 pp.

Short biographical notes on Amaldi by Conversi, F. Perain, E. Segre; personal recollections by L. W. Alvarez under the title "Recent Developments in Particle Physics" (Nobel Lecture, 11 December 1968), 1-49; articles reviewing aspects of particle physics; list of papers by Amaldi.

* Dirac, P. A. M. "Theory of electrons and positrons." Cited above as item HC24.

JF16 Dollezhal', N. A. "Rol' Sovetskikh Uchenykh v Razvitii Atomnoĭ Energetiki." *Vestnik AN SSSR*, 1 (1968): 25-34.

The role of Soviet scientists in the development of atomic energy.

JF17 Dorman, I. V. "Die Theorie Diracs und die Entdeckung des Positrons in der Kosmischen Strahlung." *NTM*, 18, no. 1 (1981): 50-57.

JF18 Dorman, I. V. "Otkrytie Mezonov." *Voprosy Istorii Estestvoznaniiâ i Tekhniki*, 1 (1982): 53-60.

Deals with the discovery of the mu and pi mesons (on the basis of published articles).

JF19 Dragoni, Giorgio. "Un momento della vita scientifica italiana degli anni trenta: la scoperta dei neutroni lenti e la loro introduzione nella sperimentazione fisica." Physis, 2 (1976): 131-164.

An interpretation of the work of Fermi's group, in particular the discovery of properties of slow neutrons.

JF20 Feather, N. "The history of atomic disintegration." Electronic Engineering, 17 (1945): 668-70.

JF21 (Fermi, Enrico) Wilson, Fred L. "Fermi's theory of Beta Decay." American Journal of Physics, 36 (1968): 1150-60.

Includes a complete English translation of Fermi's paper originally published in Zeitschrift fuer Physik (1934).

JF22 Fermi, Enrico. "Artificial radioactivity produced by neutron bombardment." Nobel Lectures...Physics (item A87), vol. 2, pp. 414-21. Delivered December 12, 1938.

JF23 Frank, I. M. "Nachalo Issledovaniĭ po Îadernoĭ Fizike v FIAN i Nekotrorye Sovremennye Problemy Stroeniîa Atomnykh Iader." Uspekhi Fizicheskikh Nauk, 91 (1967): 11-27.

The beginning of nuclear research at the Physical Institute P. N. Lebedev of AN SSSR.

JF24 Frisch, O. R. "Early steps toward the chain reaction." Rudolf Peierls and Theoretical Physics (item BP10), pp. 18-27.

* Frisch, O. R. et al. Eds. Beitraege zur Physik und Chemie des 20 Jahrhunderts. Cited above as item B11.

JF25 Grinberg, A. P. "On the history of the study of isomerism." Soviet Physics Uspekhi, 23 (1980): 848-57. Translated from Uspekhi Fizicheskikh Nauk, 132 (1980): 663-78.

The first "true case" of nuclear isomerism was discovered by O. Hahn in 1921. Early work is discussed from a modern point of view, emphasizing the work of I. V. Kurchatov's group.

JF26 Hahn, Otto. _New Atoms, Progress and Some Memories_. A collection of papers edited by W. Gaade. New York: Elsevier, 1950. 183 pp.

Contents: "From the natural transmutations of uranium to its artificial fission" (Nobel Lecture, 1946); "The chain reaction of uranium" and "Artificial new elements" (based on lectures in 1947); "Some personal reminiscences from the history of natural radioactivity" (from _Naturwissenschaften_, 1948); new sections on the hydrogen bomb and elements 97 & 98 have been added to the English edition by the author.

JF27 (Hahn, Otto) Graetzer, Hans G. "Discovery of Nuclear fission." _American Journal of Physics_, 32 (1964): 9-15.

Complete translation of the article by O. Hahn & F. Strassmann in _Naturwissenschaften_ (1939).

JF28 Hahn, Otto. "From the natural transmutations of uranium toits artificial fission." _Nobel Lectures. Chemistry_ (item A86), vol. 3, pp. 51-66. December 13, 1946. (prize for 1944)

JF29 Heilbron, J. L., Robert W. Seidel and Bruce R. Wheaton. "Lawrence and his Laboratory: Nuclear Science at Berkeley." _LBL News Magazine_, vol. 6, no. 3 (Fall 1981): 106 pp.

Special issue by Lawrence Berkeley Laboratory and Office for History of Science and Technology, University of California, Berkeley, containing "A Historian's View of the Lawrence Years."

JF30 Hess, Victor F. "Unsolved problems in pnysics: Tasks for the immediate future in cosmic ray studies." _Nobel Lectures...Physics_ (item A87), vol. 2, pp. 360-62. December 12, 1936.

JF31 Hillas, A. M. _Cosmic Rays_. New York: Pergamon Press, 1972. x + 297 pp.

Includes reprints or translatiions of papers by V. Hess (1912), W. Bothe & W. Kolhorster (1929), J. Clay (1932), C. D. Anderson (1933), J. F. Carlson & J. R. Oppenheimer (1937), S. N. Neddermeyer & C. D. Anderson (1937), P. Auger et al. (1939), C. M. G. Lattes et al. (1947), E. Fermi (1949), H. L. Bradt & B. Peters (1950), V. L. Ginzburg (1956), E. N. Parker (1958) et al.

JF32 Hinokawa, Shizue. "The history of the instrumental theory of the betatron." (in Japanese) _Kagakusi Kenkyu_, 17 (1978): 12-22.

JF33 Holloway, M. G. and C. P. Baker. "How the barn was born." Physics Today, 25, no. 7 (July 1972): 9. Reprinted from a Los Alamos report written in 1944.

JF34 Hoyer, Ulrich. "Kernphysik und Politik in der jungeren Geschichte." Kernenergie und wissenschaftliche Verantwortung. Kronberg, 1977, pp. 52-62.

JF35 Hoyer, Ulrich. "Enrico Fermi, Robert Oppenheimer, Edward Teller -- Die experimentelle Kernforschung." Die Grossen der Weltgeschichte. Edited by Kurt Fassmann et al. Bd. 11. Zurich, 1978, pp. 73-85.

JF36 Humphreys, Willard C., Jr. Anomalies and scientific theories. San Francisco: Freeman, Cooper, 1968. 318 pp.

Discusses the history of the meson, pp. 248-97.

JF37 Igonin, V. V. "The main stages and principal trends in the development of nuclear physics in the USSR." Papers by Soviet Scientists (item M48), pp. 29-45.

JF38 (Ioffe, A. F.) Vklad Akademika A. F. Ioffe v stanovlenie iadernoĭ fiziki v SSSR. (Akademiia Nauk SSSR, Ordena Lenina Fiziko-Tekhnickeskii Institut im. A. F. Ioffe Leningradskoe Otdelenie Arkhiva AN SSSR) Leningrad: Nauka, Leningradskoe Otdelenie, 1980. 38 pp. V. M. Tuchkevich & V. Ia. Frenkel, editors.

Collection of archival documents concerning Ioffe's influence on establishing Soviet nuclear physics.

JF39 Ivanenko, D. D. "Razvitie fiziki elementarhykh chastits." Voprosy Filosofii, 5 (1958): 74-87.

Development of elementary particle physics.

JF40 Johnson, Charles W. and Charles O. Jackson. City Behind a Fence. Oak Ridge, Tennessee, 1942-1946. Knoxville: University of Tennessee Press, 1981. xxiv + 248 pp.

Discusses the social history of the community organized in connection with the atomic weapons project.

JF41 Joliot, Frederic. "Chemical evidence of the transmutation of elements." Nobel Lectures...Chemistry (item A86), vol. 2, pp. 369-73. December 12, 1935.

JF42 Joliot-Curie, Irene. "Artificial production of radioactive elements." Nobel Lectures..Chemistry (item A86), vol. 2, pp. 366-68. December 12, 1935.

JF43 Jungk, Robert. Heller als tausend Sonnen. Stuttgart: Scherz & Goverts, 1956. 368pp. Translations: Brighter than a thousand suns. The Moral and Political History of the Atomic Scientists. London: Gollancz & Hart-Davis, 1958. 350pp. Gli apprendisti stregoni. Storia degli scienziati atomici. Torino: Einaudi, 1958. 380 pp.

The accuracy of this account has been challenged by many of the participants but it forms the starting point for the debate about whether German physicists would have developed the atomic bomb for Hitler.

JF44 Kedrov, B. M., Ed. Neitron: Predystoriia, Otkrytie, Posledstviia. Moskva: Nauka, 1975. 173 pp.

Neutron: prehistory, discovery, subsequent development.

JF45 Kedrov, B. M. "Slavnoe Desiatiletie Iadernoi Fiziki." Priroda, 1 (1965): 2-11.

On the "glorious" decade of nuclear physics, 1930-1940.

JF46 Kedrov, B. M. Ed. Das Neutron. Eine Artikelsammlung. Berlin: Akademie-Verlag, 1979. 269 pp.

JF47 Krafft, Fritz. "Ein fruehes Beispiel interdisziplinaerer Teamarbeit. Zur Entdeckung der Kernspaltung durch Hahn, Meitner und Strassmann." Physikalische Blaetter, 36 (1980): 85-89, 113-18.

See also the critical comment by Dietrich Hahn, "Otto Hahn und Fritz Strassmann, ibid. 37 (1981): 44-46.

JF48 Kragh, Helge. "The concept of the monopole: A historical and analytical case-study." Studies in History and Philosophy of Science, 12 (1981): 141-72.

Mostly on Dirac's 1931 theory; also discusses the "false 1975 discovery" by P. B. Price et al.

JF49 Lawrence, Ernest O. "The evolution of the cyclotron." Nobel Lectures...Physics (item A87), vol. 2, pp. 430-43.

"Owing to the war conditions, the Prize was handed over to Professor Lawrence at a ceremony in Berkeley on February 29, 1940." This lecture was delivered December 11, 1951.

JF50 Lederman, Leon. "Resource letter Neu-1. History of the Neutrino." American Journal of Physics, 38 (1970): 129-36.

Annotated bibliography.

JF51 Leprince-Ringuet, L. "Reflexions sur quarante annees de sciences nucleaires." Bulletin, Union Catholique des scientifiques francais, 126 (1972): 19-25.

JF52 Livingston, M. Stanley, Ed. The Development of High-Energy Accelerators. New York: Dover, 1966. xi + 317 pp.

Reprints of papers by J. D. Cockcroft & E. T. S. Walton (1932), R. J. Van de Graaff (1931), M. A. Tuve, L. R. Hafstad & O. Dahl (1935), H. A. Barton, D. W. Mueller & L. C. Van Atta (1932), R. G. Herb, D. B. Parkinson & D. W. Kerst (1935), G. Ising (1924), R. Wideroe (1928), E. O. Lawrence & N. E. Edlefsen (1930), E. O. Lawrence & M. S. Livingston (1932), D. H. Sloan & E. O. Lawrence (1931), L. W. Alvarez (1946), D. W. Kerst (1941), V. Veksler (1945), E. M. McMillan (1945), and others.

JF53 Livingston, Stanley. "Early history of particle accelerators." Advances in Electronics and Electron Physics, 50 (1980): 1-88.

Included in item A77. Starting with the origins of accelerators in 19th century research on electrical discharge in gases, the author gives a detailed description of work from the 1920s through the 1950s, including his own work with E. O. Lawrence. There is also a survey of accelerators constructed in the 1960s and early 1970s. 29 illustrations, 113 references.

JF54 Murphy, George M. "The discovery of deuterium." Isotopic and Cosmic Chemistry. Edited by H. Craig, S. L. Miller, and G. J. Wasserburg. Amsterdam: North-Holland, 1964, 1-7.

On Harold Urey.

JF55 Nemenov, L. M. "Istoriia razvitiia tsiklotrona za 50 let (1930-1980)." Uspekhi Fizicheskikh Nauk, 3 (1981): 525-41.

History of the cyclotron.

JF56 Petrzhak, K. A. "Kak Bylo Otkryto Spontannoe Delenie." Khimiia i Zhizn', 4 (1970): 54-58.

On the discovery (1938) by G. N. Flerov and K. A. Petrzhak of spontaneous fission.

JF57 Pontekorvo, Bruno. "Zagadochnye Neitrino." _Puti y Neznaemoe_, Sb. 3, Moskva, 1963, pp. 580-586.

"Mysterious neutrino." History of Pauli's discovery/prediction.

JF58 Rossi, Bruno. "Early days in cosmic rays." _Physics Today_, 34, no. 10 (Oct. 1981): 34-41.

Personal recollections of the period 1929-1932, when the author worked at Arcetri.

JF59 Seaborg, Glenn T. _Nuclear Milestones: A collection of Speeches_. Volume One: Builders & Discoverers. Washington, DC: U.S. Atomic Energy Commission, 1971. 138 pp.

JF60 Segre, Emilio. "Fifty years up and down a strenuous and scenic trail." _Annual Review of Nuclear and Particle Science_, 31 (1981): 1-18.

Personal recollections of his work with E. Fermi and others.

JF61 Shapley, Deborah. "Nuclear weapons history: Japan's wartime bomb projects revealed." _Science_, 199 (1978): 152-57.

See also letter by C. Weiner, _ibid_. p. 728.

JF62 Sherwin, M. J. "Niels Bohr and the atomic bomb: The scientific ideal and international politics, 1943-1944." _History of Twentieth Century Physics_ (item A140), pp. 352-69.

JF63 Snell, Arthur H. "The cyclotron group at the Metallurgical Laboratory, Chicago, 1940-1944." _American Journal of Physics_, 48 (1980): 971-78.

JF64 Stuewer, Roger H., Ed. _Nuclear Physics in Retrospect: Proceedings of a Symposium on the 1930s_. Minneapolis: University of Minnesota Press, 1979. xvi + 340 pp.

Includes papers by H. A. Bethe, E. Segre, O. R. Frisch, M. Goldhaber, E. M. McMillan, E. P. Wigner, R. Peierls, and J. A. Wheeler.

JF65 Tomonaga, S., et al. "Collected Papers on Meson Theory II. Intermediate and Strong Coupling Theories. Introduction by H. Yukawa." _Supplement of the Progress of Theoretical Physics_, no. 2 (1955). 218 pp.

Reprint of papers by S. Tomonaga (1941, 1946-47), T. Miyazima & S. Tomonaga (.1942), T. Miyazima (1943), T.

Miyazima et al. (1948), Z. Maki et al. (1953) For part I see item JF75.

* Trenn, Thaddeus J. _Transmutation: Natural and Artificial_. Cited above as item JC44.

JF66 Urey, Harold Clayton. "Some thermodynamic properties of hydrogen and deuterium." _Nobel Lectures...Chemistry_ (item A86) Vol. 2, pp. 339-54. February 14, 1935.

JF67 Walton, Ernest T. S. "The artificial production of fast particles." _Nobel Lectures...Physics_ (item A87), vol. 3, pp. 187-94. December 11, 1951.

JF68 Weart, Spencer R. "Scientists with a secret." _Physics Today_, 29, no. 2 (Feb. 1976): 23-30.

The argument on whether to publish on nuclear fission in 1939-40.

JF69 Weart, Spencer R. _Scientists in Power_. Cambridge, Mass.: Harvard University Press, 1979. xiii + 343 pp. Translation: _La Grande Aventure des Atomistes Francais_. Paris: Fayard, 1980. 394 pp.

On the French scientists working on nuclear fission in the 1930s and 1940s, and their interactions with society.

JF70 Weiner, Charles. "Institutional settings for scientific change: Episodes from the History of Nuclear Physics." _Science and Values: Patterns of Tradition and Change_. Edited by Arnold Thackray and Everett Mendelsohn. New York: Humanities Press, 1974, pp. 187-212.

JF71 Williams, Trevor I., ed. _A History of Technology_. Volume VI, _The Twentieth Century, c. 1900 to c. 1950_. Part I. Oxford: Clarendon Press, 1978. 690 pp.

Includes: Lord Hinton of Bankside, "Atomic Energy," 223-67; E. F. Newley, "The Development of the Nuclear Weapon," 268-83.

JF72 Wilson, Jane, ed. _All in Our Time: The Reminiscences of twelve nuclear pioneers_. Chicago: Bulletin of The Atomic Scientists, 1975. 236 pp.

Personal acounts by L. W. Alvarez, P. H. Abelson, M. D. Kamen, O. Frisch, H. L. Anderson, A. Wattenberg, J. Manley, R. R. Wilson, F. de Hoffman, B. McDaniel, V. F. Fitch, K. T. Bainbridge. Based on articles published in the _Bulletin of the Atomic Scientists_, 30, no. 4(April 1974) - 31, no. 6 (June 1975).

JF73 Winckler, J. R. and D. F. Hofmann. "Resource Letter CR-1 on Cosmic Rays." American Journal of Physics, 35 (1967): 2-12. Reprinted in Selected Papers on Cosmic Ray Origin Theories, Edited by Stephen Rosen. New York: Dover Pubs., 1969.

Annotated bibliography.

JF74 Wollan, E. O. "The other record of the first nuclear reactor start-up." American Journal of Physics, 48 (1980): 979-80.

JF75 Yukawa, H., et al. "Collected Papers on Meson Theory I. (Formalism and Models: Introduction by S. Tomonaga." Supplement of the Progress of Theoretical Physics, no. 1 (1955). 251 pp.

Reprint of papers by H. Yukawa (1935, 1937, 1942, 1949), H. Yukawa & S. Sakata (1937, 1939), H. Yukawa et al. (1938), M. Taketani & S. Sakata (1940), M. Kobayashi (1941, 1948), H. Tamaki (1942) and others. For Part II see item JF65.

JF76 Yukawa, Hideki. "Meson theory in its developments." Nobel Lectures ... Physics (item A87), vol. 3, pp. 128-34. December 12, 1949.

JF77 Zacharias, Peter. "Zur Entstehung des Einteilchen-Schalenmodells." Annals of Science, 28 (1972): 401-11.

From G. Beck (1927) to M. G. Mayer & J. H. D. Jensen (1949-50).

* * * * * * * * * *

For additional references see HW, sections I.c, d, e, t

JG. NUCLEAR PHYSICS AFTER 1945

JG1 Alburger, D. E. "The Search for new Isotopes at Brookhaven." A Festschrift for Maurice Goldhaber (item BG15), pp. 1-19.

* Amsel, G. "This week's Citation Classic: Amsel G. et al. Microanalysis by the direct observation of nuclear reactions using a 2 MeV Van de Graaff..." Cited above as item DG1.

JG2 Bloch, Felix. "The principle of nuclear induction." Nobel Lectures...Physics (item A87), vol. 3, pp. 203-16. December 11, 1952.

JG3 Blokhintsev, D. I. "Piat' Let Raboty Ob'edinennogo Instituta Îadernykh Issledovaniĭ." Atomnaia Energiîa, T.10, 4 (1961): 317-42.

5 years of work of the United Institute of Nuclear Research.

JG4 Brandow, Baird H. "This Week's Citation Classic: Brandow B. H. Linked-Cluster expansions for the nuclear many-body problem. Rev. Mod. Phys. 39 (1967): 771-828. Current Contents, Physical, Chemical & Earth Sciences, 21, no. 45 (Nov. 9, 1981): 16.

Brandow recalls the circumstances of writing this paper, which has been cited over 410 times since 1967.

JG5 Dailey, Benjamin P. "This Week's Citation Classic. Townes, C. H. and Dailey B. P. Determination of electronic structure of molecules from nuclear quadrupole effects. J. Chem. Phys. 17: 682-96, 1949." Current Contents, Physical, Chemical & Earth Sciences, No. 16 (April 20, 1981): 18.

Dailey recalls the circumstances of writing this paper, which has been cited over 430 times since 1961.

JG6 Dollezhal', N. A., A. K. Krasin. "Piat' Let Iadernoi Energetiki." Atomnaia Energiia, 7, no. 1 (1959): 5-10. Translation: Soviet Journal of Atomic Energy, 7, no. 1 (1960): 535-40.

On the fifth anniversary of the first nuclear power plant near Moscow (1954).

JG7 Endt, P. M. and C. van der Leun. "This Week's Citation Classic. Endt P M & van der Leun C. Energy Levels of Z = 11-21 nuclei (IV). Nucl. Phys. A 105: 1-488, 1967." *Current Contents, Physical, Chemical & Earth Sciences*, no. 5 (Feb. 2, 1981): 16.

The authors recall the circumstances of writing this paper, which has been cited over 890 times since 1967.

JG8 Ericson, Torleif E. O. "This Week's Citation Classic: Ericson T. The statistical model and nuclear level densities. *Advan. Phys.* 9: 425-511, 1960.'" *Current Contents, Engineering, Technology & Applied Sciences*, 12, no. 11 (March 16, 1981): 16.

JG9 Goncharov, V. V. "I. V. Kurchatov i Iadernye Reaktory." *Atomnaia Energiia*, 14, no. 1 (1963): 10-17.

On I. V. Kurchatov and nuclear reactors.

JG10 Hofstadter, Robert. "The electron scattering method and its application to the structure of nuclei and nucleons." *Nobel Lectures.Physics* (item A87), vol. 3, pp. 560-81. December 11, 1961.

JG11 Iwadare, J., et al. "Meson Theory III. Nuclear Forces. Introduction by M. Takeani." *Supplement of the Progress of Theoretical Physics*, no. 3 (1956). 174 pp.

Review-historical articles on the pion theory of nuclear forces by J. Iwadare et al., S. Machida & T. Toyoda, K. Nishijima, M. Taketani et al. For parts I and II of this series, see items JF75 and JF65.

JG12 Jensen, J. H. D. "Zur Geschichte der Theorie des Atomkerns (Nobelvortrag)." *Angewandte Chemie*, 76 (1964): 69-75. Translation: "The history of the theory of structure of the atomic nucleus." *Science*, 147 (1965): 1419-23.

JG13 Kipphardt, Heinar. *In the Matter of J. Robert Oppenheimer. A Play freely adapted on the basis of the documents*. Translated from the German edition of 1964 by Ruth Spiers. New York: Hill and Wang, 1968. 128 pp.

JG14 Kuo, T. T. S. "This Weeks' Citation Classic: Kuo T. T. S. & Brown, G. E. Structure of finite nuclei and the free nucleon-nucleon interaction: an application to ^{18}O and ^{18}F. Nuclear Phys. 85: 40-86, 1966." *Current Contents, Physical, Chemical & Earth Sciences*, 20, no. 1 (January 7, 1980): 10.

The author recalls the circumstances of writing this paper which has been cited more than 545 times since 1966.

JG15 McMillan, Edwin M. "The transuranium elements: early history." _Nobel Lectures...Chemistry_ (item A86), vol. 3, pp. 314-22. December 12, 1951.

* Maglich, Bogdan, Ed. _Adventures in Experimental Physics_. Cited above as item A74.

JG16 Martalogu, N. "Aspects du developpement de la science et des techniques nucleaires en Roumanie." _Actes du XIII Congres International d'Histoire des Sciences_. Moscow: Nauka, 1974, Sections IA, II, pp. 231-6.

Discusses planned development of nuclear energy in Roumania since 1955.

JG17 Nemirovsky, P. E. "K istorii obosnovaniia sistematiki isotopov." _Voprosy Istorii Estestvoznaniia i Tekhniki_, 1, no. 34 (1971): 20-30. English summary, p. 108.

"On the history of the establishment of the system of isotopes." Studies in nuclear physics by W. Elsasser and M. G. Mayer relating properties of nuclei to numbers of neutrons and protons.

JG18 Newman, Steven L. _The Oppenheimer Case: A reconsideration of the role of the Defense Department and national security_. Ph.D. Dissertation, New York University, 1977. 202 pp.

For summary see _Dissertation Abstracts International_, 38 (1977): 2306-A.

JG19 Purcell, Edward M. "Research in nuclear magnetism." _Nobel Lectures...Physics_ (item A87), vol. 3, pp. 219-31. December 11, 1952.

JG20 Rainwater, James. "Background for the spheroidal nuclear model proposal." _Science_, 193 (1976): 378-83.

JG21 Seaborg, Glenn T. "The periodic table: Tortuous path to man-made elements." _Chemical and Engineering News_, 57, no. 16 (1969): 46-52.

JG22 Seaborg, Glenn T. "The transuranium elements: Present status." _Nobel Lectures...Chemistry_ (item A86), vol. 3, pp. 325-49. December 12, 1951.

JG23 Segrè, Emilio. "Antinucleons." _American Journal of Physics_, 6 (1957): 363-69.

JG24 Shpinel', V. S. and A. A. Sorokin. "Iadernaia spektroskopiia." Razvitie Fiziki v Rossii (item A93), pp. 359-81.

Nuclear spectroscopy.

JG25 Smirenkin, G. N. "Delenie iader." Razvitie Fizike v Rossii (item A93), pp. 334-59.

Fission of nuclei.

* Strachan, Charles. The Theory of Beta-Decay. Cited below as item JH58.

JG26 Teller, Edward. "The work of many people." Science, 121 (1955): 267-74.

On the development of the hydrogen bomb.

* Trenn, Thaddeus J. "Thoruranium (U-236) as the extinct natural parent of thorium: the premature falsification of an essentially correct theory." Cited above as item JC45.

* Weinberg, Alvin M. Reflections on Big Science. Cited above as item C115.

JG27 Willmott, J. C. "Nuclear structure." Physics Bulletin, 19 (1968): 289-96.

* Zacharias, Peter. "Zur Entstehung des Einteilchen-Schalen-modells." Cited above as item JF77.

JH. ELEMENTARY PARTICLES, HIGH ENERGY PHYSICS

* Akchurin, I. A. "Razvitie fiziki antichatits." Cited above as item JF1.

JH1 Amaldi, Ugo. "Particle Accelerators and Scientific Culture." *Scientific Culture* (item A78), pp. 41-129.

JH2 Artsimovich, L. A. "Issledovaniia po Upravliaemym Termoiadernym Reaktsiiam v SSSR." *Vestnik AN SSSR*, 1 (1959): 11-23.

JH3 Bernardini, C. "The Story of AdA." *Scientia*, 113 (1978): 39-44. Translated from Italian version on pp. 27-38.

History of "*A*nelli *d*i *A*ccumulazione" (electron-positron storage rings), starting with Bruno Touschek's 1959 seminar at Frascati, which pointed the way to "the Italian path to high energy physics."

JH4 Blau, Judith R. "Sociometric structure of a scientific discipline." *Research in Sociology of Knowledge, Sciences, and Art*, 1 (1978): 191-206

Discusses communication networks of U. S. high-energy physicists, and notes tendency for most creative contributions to be made by young physicists because of the social structure of the field.

JH5 Brush, Stephen G. "The scientific value of high energy physics." *Annals of Nuclear Energy*, 8 (1981): 133-40.

On methods for estimating the cost and significance of discoveries in different areas of physics, and on the argument that high energy research is "fundamental" to science, with historical perspective on that concept.

JH6 Chamberlain, Owen. "The early antiproton work." *Nobel Lectures...Physics* (item A87), vol. 3, pp.489-505. December 11, 1959.

JH7 Chang, N. P., Ed. "Five decades of weak interactions." *Annals of the New York Academy of Sciences*, 294 (1977). 102 pp.

Symposium in honor of R. E. Marshak, with papers by Y. Ne'eman, C. S. Wu, H. A. Bethe and others.

JH8 Cini, Marcello. "The history and ideology of dispersion relations. The pattern of internal and external factors in a paradigmatic shift." *Fundamenta Scientiae*, 1 (1980): 157-72.

The popularity of this theoretical technique for interpreting elementary particle phenomena in the late 1950s is attributed to the role of the American physics community which was markedly different from that of the European physics community in 1925 when the relations were first proposed. The papers by M. L. Goldberger (1955), G. F. Chew (1958) and others are discussed.

* Conversi, M., ed. *Evolution of Particle Physics: A Volume dedicated to Edoardo Amaldi in his 60th Birthday*. Cited above as item JF15.

JH9 Cronin, James W. " CP symmetry violation: the search for its origin." *Science*, 212 (1981): 1221-28.

Nobel lecture.

JH10 Cronin, James W., and Margaret Stautberg Greenwood. "CP symmetry violation." *Physics Today*, 35, no. 7 (July 1982): 38-44.

"In an informal discussion that grew out of a recent talk to physics teachers in Chicago, the codiscoverer of CP asymmetry recalls the circumstances of the observation and discusses its implications."

JH11 Day, J. S., A. D. Krisch, and L. G. Ratner, Eds. *History of the ZGS (Argonne, 1979)*. Proceedings of the Symposium on the History of the Zero Gradient Synchrotron. New York: American Institute of Physics, 1980. 453 pp.

Includes papers by J. J. Livingood, L. C. Teng, A. V. Crewe and others.

JH12 Demy, Nicholas. "Quark etymology." *Science*, 184 (1974): 1327.

JH13 DeWitt, Hugh E. "The experience of a nuclear weapons lab physicist in the *Progressive* case." *Physics and Society*, 9, no. 3 (Oct. 1980): 11-13.

On the attempt to suppress publication of an article about the construction of a hydrogen bomb.

JH14 Drell, Sidney P. "The Richtmyer Memorial Lecture -- When is a Particle?" *American Journal of Physics*, 46 (1978): 597-606.

The problem of accepting quarks as particles is discussed in the light of the history of the neutrino and earlier cases.

JH15 Epstein, Edmund L. (Letter to the Editor). *Scientific American*, 219, no. 1 (July 1968): 8.

On the origin of word "quark" in Goethe's *Faust*.

JH16 Estulin, I. V. "Neitronnaia fizika." *Razvitie fizike v Rossii* (item A93), pp. 305-34.

Neutron physics.

JH17 Faucher, Guy. "Quark History." *Physics Today*, 34, no. 3 (March 1981): 80.

Comments on Witten's earlier article on quarks; reply by Witten, *ibid*., pp. 80-81.

JH18 Franklin, Allan. "Justification of a 'crucial' experiment: Parity nonconservation." *American Journal of Physics*, 49 (1981): 109-112.

JH19 Franklin, A. D. "What makes a 'good' experiment?" *British Journal for the Philosophy of Science*, 32 (1981): 367-79.

Discusses some examples from 20th century particle physics.

JH20 Freundlich, Yehudah. "Theory evaluation and the bootstrap hypothesis." *Studies in History and Philosophy of Science*, 11 (1980): 267-77.

The rise and fall of the hypothesis proposed by G. Chew in high energy physics.

JH21 Gaillard, Mary K. "This Week's Citation Classic. Gaillard M K, Lee B W & Rosner J L. Search for Charm. *Rev. Mod. Phys*. 47: 277-310, 1975." *Current Contents, Physical, Chemical & Earth Sciences*, no. 10 (March 9, 1981): 18.

Gaillard recalls the circumstances leading to the writing of this paper which has been cited over 540 times since 1975.

JH22 Gale, George. "Forces and particles: Concepts again in conflict." *Journal of College Science Teaching*, 3

(1973): 29-35.

JH23 Gale, George. "Chew's monadology." Journal of the History of Ideas, 35 (1974): 339-48.

JH24 Gaston, Jerry. Originality and competition in science: A study of the British High Energy Physics Community. Chicago: University of Chicago Press, 1973. 210 pp.

JH25 Gavroglu, K. "Research guiding principles in modern physics: Case studies in elementary particle physics." Zeitschrift fuer allgemeine Wissenschaftstheorie, 7 (1976): 223-48.

JH26 Glaser, Donald A. "Elementary particles and bubble chambers." Nobel Lectures...Physics (item A87), vol. 3, pp. 529-31. December 12, 1960.

JH27 Glashow, Sheldon Lee. "Towards a unified theory: Threads in a tapestry." Science, 210 (1980): 1319-23.

Nobel lecture.

JH28 Glashow, Sheldon Lee. "This Week's Citation Classic: Glashow, S L, Iliopoulos J & Maiani L. Weak interactions with lepton-hadron symmetry. Phys. Rev. D 2: 1285-92, 1970." Current Contents, Physical, Chemical & Earth Sciences, 20, no. 20 (May 19, 1980): 10.

The author recalls the circumstances of writing this paper, which has been cited more than 1085 times since 1970.

* Greenberg, Daniel S. The Politics of Pure Science. Cited above as item C44.

JH29 Hendrick, R. E. and Anthony Murphy. "Atomism and the illusion of crisis: The danger of applying Kuhnian paradigms to current particle physics." Philosophy of Science, 48 (1981): 454-68.

A critique of the paper by K. Shrader-Frechette (see below, item JH 56).

* Hillas, A. M. Cosmic Rays. Cited above as item JF31.

* Hofstadter, Robert. "The electron-scattering method and its application to the structure of nuclei and nucleons." Cited above as item JG10.

JH30 Hung, P. Q. and C. Quigg. "Intermediate bosons: weak interaction carriers." *Science*, 210 (1980): 1205-11.

* Ivanenko, D. D. "Razvitie fiziki elementarnykh chastits." Cited above as item JF39.

JH31 Jachim, A. G. *Science Policy making in the United States and the Batavia accelerator*. Carbondale: Southern Illinois University Press, 1975. 208 pp.

JH32 Jacob, M. Ed. *CERN - 25 Years of Physics*. Amsterdam & New York: North-Holland Pub. Co., 1981. vii + 560 pp.

"The purpose of this book is to provide a detailed yet convenient review of the important physics results obtained at CERN since its inception in 1954, and also to provide a thorough and up to date review of several fields of research in which CERN has played the leading role." Includes "Highlights of 25 years of physics," by L. Van Hove and M. Jacob, and specialized articles on stochastic cooling, on-line analysis of radioactive ions, very high energy hadron intereactions as studied with the CERN interacting storage rings, the g-2 experiments and neutrino physics as studied with the Gargamelle Bubble Chamber.

JH33 Jungk, Robert. *Die Grosse Machine. Auf dem Weg in eine andere Welt*. Bern: Scherz, 1966. 270 pp. New Edition. Muenchen: Deutscher Taschenbuch-Verlag, 1969. 282 pp. Translations: *La grande macchina. I nuovi scienziati atomici*. Torino: Einaudi, 1968. 244 pp. *The Big Machine*. New York: Scribner, 1968. vii + 245 pp.

On accelerators and high energy physics research.

JH34 Kolman, E. "Die Dialektik der Entwicklung der moderne Kernphysik." *NTM, Schriftenreihe fuer Geschichte der Naturwissenschaft, Technik und Medizin*, 4, no. 9 (1967): 42-45.

Marxist view of recent particle physics.

JH35 Kuti, Julius. "This week's citation classic: Kuti, J. and V. F. Weisskopf. Inelastic lepton-nucleon scattering and lepton pair production in the relativistic quark-parton model. Phys. Rev. D4: 3418-39, 1971." Current Contents, Physical, Chemical & Earth Sciences, 20, no. 52 (29 Dec. 1980): 12.

Kuti recalls the circumstances under which he and Weisskopf originally wrote this paper, which has been cited over 300 times. "To me, the major importance of this paper lies in the fact that it was the first to show to theorists and experimentalists that a reasonably simple quark-parton model of the nucleon can explain a vast amount of observations, provide us with new testable predictions, and strengthen our belief in the quark and gluon constituents as new building blocks of nature below the known nuclear scale."

* Kragh, Helge. "The concept of the monopole: A historical and analytical case-study." Cited above as item JF49.

JH36 Lee, T. D. "Weak interactions and nonconservation of parity." Nobel Lectures...Physics (item A87), vol. 3, pp. 406-418. December 11, 1957.

* Livingston, M. Stanley, Ed. The Development of High-Energy Accelerators. Cited above as item JF53.

Includes reprints of papers by D. Bohm and L. Foldy (1946-47), M. S. Livingston et al. (1950), E. D. Courant, M. S. Livingston & H. S. Snyder (1952), N. Christofilos (1956) and others.

* Maglich, Bogdan, Ed. Adventures in Experimental Physics. Cited above as item A74.

JH37 Moravcsik, M. J. "The crisis in particle physics." Research Policy, 6 (1977): 78-107.

See comment by Polkinghorne, item JH 46.

JH38 Ne'eman, Yuval. "Concrete versus abstract theoretical models." The Interaction between Science and Philosophy. Edited by Y. Elkana. Atlantic Highlands, NJ: Humanities Press: 1974, pp. 1-25.

Comments on the history of particle theory, including Marxist influences.

* Ne'eman, Yuval. "A view from the bridge: Physics and philosophy in the 20th century." Cited below as item

K77.

JH39 Okubo, Susumu. "This Week's Citation Classic: Okubo, S. Note on unitary symmetry in strong interactions. _Prog. Theor. Phys_. 27: 949-66, 1962." _Current Contents, Physical, Chemical, and Earth Sciences_, 20, no. 36 (September 8, 1980): 16.

The author recalls the circumstances of writing this paper, which has been cited more than 495 times since 1962.

JH40 Petrosyants, A. M. and V. P. Dzhelepov. "Uspekhi tekhniki uskoritelei elementarnykh chastits v Sovetskom Soiuze." _Maria Sklodowska-Curie: Centenary Lectures_ (item BC28), pp. 165-176. For English translations see next item.

Survey of Soviet work in accelerator technique from early efforts in the 30s by L. V. Mysovsky, K. D. Sinel'nikov and A. K. Valter' to Dubna and Serpukhov.

JH41 Petrosyants, A. M. and V. P. Dzhelepov. "Advances in the Development of elementary particles accelerators in the Soviet Union." _Maria Sklodowska-Curie: Centenary Lectures_ (item BC28), pp. 177-184.

JH42 Petržílka, Václav. "Třicet let československý experimentálni fyziky elementárnich částic." _Československý časopis pro fyziku_, sekce A, 22 (1972): 426-28.

Thirty years of Czechoslovak experimental physics of elementary particles.

JH43 Pickering, Andrew. "The role of interests in high-energy physics. The choice between charm and colour." _The Social Process of Scientific Investigation_ (Sociology of the Sciences, Volume IV). Edited by K. D. Knorr, R. Krohn and R. Whitley. Boston: Reidel, 1980, pp. 107-38.

Influence of discovery of J-psi particle; work of G. Feldman & P. Matthews (colour), S. L. Glashow et al. (charm).

JH44 Pickering, Andrew. "Exemplars and analogies: A Comment on Crane's study of Kuhnian paradigms in high energy physics." _Social Studies of Science_, 10 (1980): 497-502.

With Crane's "Reply to Pickering," _ibid_. 502-6 and Pickering's "Reply to Crane," _ibid_. 507-8. Comments on theories of S. Weinberg, A. Salam, S. Glashow.

JH45 Pickering, Andrew. "The Hunting of the Quark." Isis, 72 (1981): 216-36.

On the experiments at Genoa (led by G. Morpurgo) and at Stanford (led by W. M. Fairbank).

JH46 Pickering, Andrew. "Constraints on Controversy: The case of the Magnetic Monopole." Social Studies of Science, 11 (1981): 63-93.

On the 1975 discovery claim by P. B. Price, E. K. Shirk, W. Z. Osborne and L. S. Pinsky.

JH47 Polkinghorne, J. C. "Particle physics - an alternative view." Research Policy, 6 (1977): 412-15.

Comments on Moravcsik (item JH37)

JH48 Powell, Cecil F. "The cosmic radiation." Nobel Lectures...Physics (item A87) vol. 3, pp. 144-57. December 11, 1950.

JH49 Richter, Burton. "From the Psi to Charm: The experiments of 1975 and 1976." Science, 196 (1977): 1286-97.

Nobel lecture.

JH50 Rosner, Jonathan L. "Resource Letter NP-1: New particles." American Journal of Physics, 48 (1980): 90-103. Reprinted in New Particles... (item JH51).

Annotated bibliography on the growth of particle physics since 1974.

JH51 Rosner, Jonathan L. Ed. New particles: Selected Reprints. Stony Brook, NY: American Association of Physics Teachers, 1981. 121 pp.

Includes item JH50 and reprints 25 articles on the J/psi particle, charmed particles, tau and upsilon leptons, etc.

JH52 Sakata, Shoichi. (50 years of the quantum theory. 2. Theory of elementary particles. -- in Japanese). Journal of History of Science, Japan, no. 21 (1952): 5-9.

JH53 Salam, Abdus. "Progress in Renormalization Theory since 1949." The Physicist's Conception of Nature: (item A80), pp. 430-46.

JH54 Sciulli, F. "An experimenter's history of neutral currents." Progress in Particle and Nuclear Physics, 2 (1979): 41-87.

JH55 Segre, Emilio G. "Properties of antinucleons." Nobel Lectures...Physics (item A87), vol. 3, pp. 508-20. December 11, 1959.

* Segre, Emilio. "Fifty years up and down a strenuous and scenic trail." Cited above as item JF60.

JH56 Shrader-Frechette, K. "Atomism in crisis: An analysis of the current high-energy paradigm." Philosophy of Science, 44 (1977): 409-40.

Uses T. S. Kuhn's theory of scientific revolutions to argue that recent developments "presage a conceptual revolution" in high energy physics. See critique by Hendrick and Murphy (item JH29).

JH57 Shull, C. G. "Physics with Early Neutrons." Proceedings of the Conference on Neutron Scattering, Pt. I. Springfield, VA: US Dept. of Commerce, 1976, pp. 1-16.

Preliminary paper of a conference held at Gatlinburg, TN, 6-10 June 1976. Account of early development in neutron scattering research from the beginning of 1946 by Oak Ridge group utilizing Clinton Pile.

JH58 Strachan, Charles. The theory of Beta Decay. New York: Pergamon Press, 1969. viii + 213 pp.

Includes reprints of papers by E. Fermi (1934), C. L. Cowan, F. Reines, F. B. Harrison, H. W. Kruse and A. D. McGuire (1956), G. Gamow and E. Teller (1936), T. D. Lee and C.N. Yang (1956), C. S. Wu, E. Ambler, R. W. Hayward, D. D. Hoppes and R. P. Hudson (1957), T. D. Lee & C. N. Yang (1957), M. Goldhaber, L. Grodzins & A. W. Sunyar (1958), R. P. Feynman & M. Gell-Mann (1958), and M. Gell-Mann (1958).

JH59 Stwertka, A. and E. M. Stwertka. "The devil's quark." New York Times, 16 January 1976, p. 28.

Origin of the word "quark" in Goethe's Faust.

JH60 Sullivan, Daniel, D. H. White, and E. J. Barboni. "The State of a Science: Indicators of the Speciality of Weak Interactions." Social Studies of Science, 7 (1977): 167-200.

A statistical study of the literature, 1950-72.

JH61 Swetman, T. P. "The response to crisis -- A contemporary case study." American Journal of Physics, 39 (1971): 1320-28.

Concludes that T. S. Kuhn's paradigm theory applies to the 1964 discovery of the decay of K_2^0 (time reversal symmetry problem).

JH62 Taylor, Hugh. "From 'The Lynxes' to Stanford's 'Lin-Ac.'" American Scientist, 54 (1966): 333-44.

Development of particle accelerators in the context of ideas about atoms.

JH63 Telegdi, V. L. "Crucial experiments on Discrete Symmetries." The Physicist's Conception of Nature (item A80), pp. 454-80.

JH64 Ting, Samuel, C. C. "The discovery of the J particle: a personal recollection." American Scientist, 65 (1977): 1167-77.

* Tomonaga, S., et al. "Collected Papers on Meson Theory II." Cited above as item JF65.

* Weinberg, Alvin M. Reflections on Big Science. Cited

JH65 Weinberg, Steven. "Conceptual foundations of the unified theory of weak and electromagnetic interactions." Science 210 (1980): 1212-18.

Nobel lecture.

JH66 Wideröe, Rolf. "Das Betatron." Zeitschrift fuer angewandte Physik, 5 (1953): 187-200.

Includes historical remarks on the work of J. Slepian (U. S. patent, 1922) and R. Wideroe (1922-27, pub. 1928); deals mostly with developments since 1945.

JH67 Wilson, R. R. "US Particle Accelerators at age 50." Physics Today, 34, no. 11 (Nov. 1981): 86-103.

Survey for the special issue "50 years of physics in America."

JH68 Yang, Chen Ning. "The law of parity conservation and other symmetry laws of physics." Nobel Lectures...Physics (item A87), vol. 3, pp. 393-403.

* * * * *

For additional references see HW, Section K.

K. PHYSICS AND PHILOSOPHY

K1 Barker, Peter. "Hertz and Wittgenstein." Studies in History and Philosophy of Science, 11 (1980): 243-56.

Discusses the influence of H. Hertz (primarily his writings on mechanics) on the Tractatus Logico-Philosophicus of L. Wittgenstein.

K2 Bavink, Bernhard. Ergebnisse und Probleme der Naturwissenschaften. Eine Einfuehrung in die heutige Naturphilosophie. 6th ed. Leipzig: Hirzel, 1940. 796 pp. Translation: Risultati e Problemi delle Scienze Naturali. Introduzione alla Filosofia naturale dei nostri giorni (Translated from the 8th German edition by E. Colorni, ed. by M. Ageno) Firenze: Sansoni, 1948. 320 pp.

K3 Bergmann, Hugo. Der Kampf um das Kausalgesetz in der juengsten Physik. Braunschweig: Vieweg & Sohn, 1929. 78 pp. Translation, "The controversy concerning the law of causality in contemporary physics." Boston Studies in the Philosophy of Science, 13 (1974): 395-462.

K4 Bohm, David. Causality and Chance in Modern Physics. London: Routledge & Kegan Paul, 1957; reprint, New York: Harper, 1961. xi + 170 pp.

A thoughtful but readable discussion of the philosophical issues involved in classical and quantum physics. Bohm suggests but does not elaborate here his own alternative interpretation of quantum theory.

K5 Bohr, Niels. Atomic Theory and the Description of Nature. Cambridge: Cambridge University Press, 1934, 1961. 119 pp.

Reprints several popular lectures and articles, including his earliest formulation of the "Copenhagen Interpretation" of quantum mechanics.

K6 Bohr, Niels. Atomic Physics and Human Knowledge. New York: Wiley, 1958. Reprint, New York: Science Editions, 1961. viii + 101 pp.

Includes his 1933 address "Light and Life" and "Discussion with Einstein on Epistemological Problems in Atomic Physics," along with other lectures on physics, biology, and the unity of knowledge.

K7 Bohr, Niels. I quanti e la vita. Torino: Boringhieri, 1961.

K8 Bohr, Niels. *Teoria dell'atomo e conoscenza umana.* Torino: Boringhieri, 1961. 449 pp.

Translation of "Light and Life" from item K6.

K9 Bohr, Niels. *Essays 1958-1962 on Atomic Physics and Human Knowledge.* New York: Wiley, 1963. Reprint, New York: Vintage Books, 1966. x + 100 pp.

A collection of lectures given on various occasions; includes reminiscences of E. Rutherford, "The genesis of quantum mechanics," and "The Solvay meetings and the development of quantum physics."

K10 [Bohr, N.] B.Blazek. "Bohr a Piaget: Poznani jako otevreny cirkularni proces." *Ceskoslovensky Casopis pro Fyziku*, sekce A, 25 (1975): 170-76.

K11 [Bohr, N.] Favrholdt, David. "Niels Bohr and Danish Philosophy." *Danish Yearbook of Philosophy*, 13 (1976): 206-20.

K12 [Bohr, N.] Honner, John. "The transcendental philosophy of Niels Bohr." *Studies in History and Philosophy of Science*, 13 (1982): 1-29.

K13 Bondi, Hermann. *Assumption and Myth in Physical Theory*. London: Cambridge University Press, 1967. 85 pp.

K14 Bondi, Hermann. *Miti e ipotesi nella teoria fisica. Demistificazione di alcune grandi idee della fisica.* Bologna: Zanichelli, 1971. 102 pp.

K15 Born, Max. *Natural Philosophy of Cause and Chance*. Oxford: Clarendon Press, 1949. viii + 215 pp. Translation: *Filosofia naturale della causalita e del caso*. Torino: Boringhieri, 1962. 266 pp.

While including brief statements of the author's "metaphysical conclusions," the book consists mainly of an exposition of kinetic theory and statistical mechanics.

K16 Bridgman, Percy Williams. *The Logic of Modern Physics*. New York: Macmillan, 1927. xiv + 228 pp. Translation: *Die Logik der heutigen Physik*. Muenchen: Max Hueber Verlag, 1932. xii + 170 pp. *La Logica dell fisica moderna*. Torino: Einaudi, 1952. 204 pp.

K17 Bridgman, P.W. *The Nature of Physical Theory*. Princeton, N. J.: Princeton University Press, 1936. Reprint, New York: Dover Pubs., n. d. 138 pp. Translation: *La natura della teoria fisica*. Firenze: La Nuova Italia, 1965. 163 pp.

K18 Bridgman, P. W. La Critica operazionale della scienza. Torino: Boringhieri, 1969. 454 pp.

K19 Brillouin, Leon. Vie, Matiere et Observation, Editions Albin Michel, Paris, 1959. 239 pp.

On thermodynamics, information theory, determinism, Poincare, Einstein.

K20 Broda, E. "Boltzmann, Einstein, Natural Law and Evolution." Comparative Biochemistry and Physiology, 67B (1980): 373-78.

Boltzmann's writings reflect the influence of an evolutionary viewpoint but Einstein's do not.

K21 Broglie, Louis de. Physique et Microphysique, Albin Paris: Michel, 1947. 370 pp. Translation: Fisica e microfisica. Torino: 1950. 347 pp.

K22 Brouzenc, Paul. "Magnetisme et energetique: La methode de Duhem. A propos d'une lettre inedite de Pierre Curie." Revue d'Histoire des Sciences, 31 (1978): 333-44.

K23 Brunschvicg, Leon. La Physique du Vingtieme Siecle et La Philosophie. Paris: Hermann, 1936. 31 pp.

K24 Brush, Stephen G. "Can science come out of the laboratory now?" Bulletin of the Atomic Scientists, 32, no. 4 (April 1976): 40-43.

* Brush, Stephen G. "Statistical Mechanics and the Philosophy of Science." Cited above as item DB10.

K25 Bruzzaniti Giuseppe. "'Real History' as 'Dictionary' Reconstruction." Scientia, 115 (1980): 643-61.

"A Historiographic Hypothesis for Pierre Curie's Scientific Undertaking."

K26 Bunge, Mario. Causality;: The Place of the Causal Principle in Modern Science. Cambridge, Mass.: Harvard University Press, 1963. Translation: La Causalita. Il posto del principio causale nella scienza moderna. Torino: Boringhieri, 1970. 468 pp.

K27 Capek, Milic. *The Philosophical Impact of Contemporary Physics*. Princeton, N. J.: Van Nostrand, 1961. xvii + 414 pp.

Reviews the classical concepts of space, time, matter motion, etc. and their transformation in the 20th century; concludes with chapters on "The End of the Laplacian Illusion [of determinism]" and "The Reinstatement of Becoming in the Physical World" (novelty and irreversibility).

* Cassirer, E. *Determinismus und Indeterminismus in der modernen Physik*. Cited below as item HD13.

K28 Chandrasekhar, S. "Beauty and the quest for beauty in science." *Physics Today*, 32, no. 7 (July 1979): 25-30.

K29 Cohen, Robert S. and Raymond J. Seeger, eds. *Ernst Mach: Physicist and Philosopher*. (Boston Studies in the Philosophy of Science, 6) New York: Humanities/ Dordrecht: Reidel, 1970. viii + 295 pp.

Includes articles by P. G. Bergmann, G. Holton, H. Goenner, P. Frank, R. von Mises and others on Mach's influence on 20th century physics.

K30 Delokarov, K. Kh. *Osobennosti Stanovleniîa Soiuza Filosofii i Fiziki v SSSR. 20-e-Nachalo 30-Kh Godov*. Moskva: Izdatel'stvo Moskovskiĭ Gosudarstvennyĭ Universitet im. M.V. Lomonosova, 1969. 26 pp. (Nauchnaya Konferentsiîa "Leninskiĭ Etap v Razvitii Marksistskoi Filosofiĭ." Leningrad, 16-19 Dek. 1969).

The peculiarity of the union of physics and philosophy in USSR in the 20s and early 30s.

K31 Delokarov, K. Kh. "Iz istorii bor'by za utvershdenie dialekticheskogo materializma v Sovetskoĭ fizike, 1920-1930." *Uchenye Zapiski Moskovskii Gosudarstvennyĭ Pedagogicheskii Institut im V. I. Lenina*, 290 (1968): 5-20.

From the history of affirmation of dialectical materialism in Soviet physics, 1920-1930.

K32 Diederich, Mary E. "The context of inquiry in physics." *American Journal of Physics*, 40 (1972): 449-57.

The process of theory evaluation, illustrated by the work of N. Bohr, M. Planck, and A. Einstein.

K33 Dingle, Herbert. "Philosophy of physics, 1850-1950." *Nature*, 168 (1951): 630-36.

K34 [Dirac, P.] Kragh, Helge. Methodology and Philosophy of Science in Paul Dirac's Physics. Roskilde, Denmark: Roskilde Universitetscenter, IMFUFA, Tekst Nr. 27, 1979. 139 pp.

Concentrates on three aspects of the work of P. A. M. Dirac: relativistic quantum theory, the theory of magnetic monopoles, and theory of cosmological constants. Concludes that the principle of plentitude and the principle of beauty were important in his career.

K35 Duhem, Pierre. La theorie physique: son objet et sa structure. Paris: Chevalier and Riviere, 1906. 450 pp. Translations: The aim and structure of physical theory. Princeton, NJ: Princeton University Press, 1954. xii + 344 pp. La Teoria fisica e la sua struttura. Bologna: II Mulino, 1978. xvi + 386pp.

K36 Du Plessis, S. I. M. The Compatibility of Science and Philosophy in France, 1840-1940. Cape Town: Balkema, 1972. 300 pp.

Chapter IV, "The Status of Physics" (pp. 101-59) discusses the views of H. Poincare, P. Duhem, E. Meyerson, L. Brunschvicg, E. Boutroux, H. Bergson and others.

K37 Einstein, A. Mein Weltbild. Amsterdam: Querido verlag, 1934. 269 pp. Translations: The world as I see it. Alan Harris, trans. New York: Covici, Friede, 1934. 200 pp. Come vedo il mondo. Giachini: Milano, 1952. 151 pp.

K38 Einstein, A. Pensieri degli anni difficili, Torino: Boringhieri, 1965.

K39 [Einstein, A.] Chudinov, E. M., ed. Ĕinshteĭn i Filosofskie Problemy Fiziki XX Veka. Moskva, 1979. 568 pp.

19 papers on Einstein and philosophical problems of 20th century physics by: V. A. Fok, A. D. Alexandrov, V. L. Ginzburg, M. A. Markov, V. S. Barashenkov, V. I. Rodichev, D. P. Pribanov, M. E. Omel'îanovskii, B. G. Kuznetsov, I. A. Akchurin and M.D. Akhundov, G. E. Gorelik, A. M. Mostepanenko, E. M. Chudinov, K. Kh. Delokarov, etc.

K40 [Einstein, A.] Gribanov, D. P. "The philosophical views of Albert Einstein." Soviet Studies in Philosophy, 18, no. 2 (1979): 72-93. Translated from Voprosy Filosofii, 1979.

K41 [Einstein, A.] Janich, Peter. "Die Erkenntnistheoretischen Quellen Einsteins." _Einstein Symposium, Berlin_ (item BE68), pp. 412-27.

K42 [Einstein, A.] Polikarov, Azaria. "Einstein's Conception of Physical Theory," _Epistemologia_, 1 (1979): 99-120.

K43 Enriques, Federigo. _Causalita e determinismo nella filosofia e nella storia della scienza_. Roma: Atlantica Editrice. 109 pp.

K44 Enriques, Federigo. _Problemi della Scienza_. Bologna: Zanichelli, 1906. 593 pp.

K45 Feuer, L. S. "Teleological principles in science." _Inquiry_, 21 (1978): 377-406.

Includes Gell-Mann's principle, in the search for elementary particles.

K46 Feyerabend, Paul. "Zahar on Mach, Einstein and Modern Science." _British Journal for the Philosophy of Science_, 31 (1980): 273-82.

Critique of Zahar's paper, item K110.

K47 Frank, Philipp. _Die Kausalgesetz und seine grenze_. Wien: J. Springer, 1932. xv + 308 pp.

K48 Frank, Philipp. _Modern Science and its Philosophy_. Cambridge: Harvard University Press, 1949. 316 pp. Translation: _La Scienza moderna e la sua filosofia_. Bologna: Il Mulino, 1973. 336 pp.

Essays on E. Mach, quantum theory, indeterminism, logical empiricism, and other philosophical aspects of physics.

K49 Gale, George. "Leibniz, Chew, and Wheeler on the identity of physical and philosophical inquiry." _Review of Metaphysics_, 29 (1975): 322-33.

K50 Garbasso, Antonio. _Fisica d'oggi Filosofia di domani_. Milano: Libreria Editrice Milanese, 1910. 192pp.

K51 Graham, Loren R. _Between Science and Values_. New York: Columbia University Press, 1981. x + 449 pp.

"A discussion of the relationship between science and values in the 20th century, concentrating on the impact of physics and biology on conceptions of social values. The author studies the views of a number of "Authoritative popularizers of science" (e.g. Einstein, Bohr, Heisenberg, Eddington, Monod, Skinner, Lorenz, Wilson) and classifies them either as 'Expansionists' or 'Restrictionists. (The former) believe that science can be related, either directly or indirectly, to value questions; (the latter) contend that science and values belong to separate realms. The author...concludes that Restrictionism is dead as an intellectual option...(but that) Expansionism...must be subjected to rigorous criticism in order to avoid the abuse of science for political purposes." (Summary provided by author).

K52 Gunter, Peter A. Y., ed. Bergson and the Evolution of Physics. Knoxville: University of Tennessee Press, 1969. xi + 348pp.

Includes articles by P. A. Y. Gunter, L. de Broglie, S. Watanabe, O. Costa de Beauregard, R. Blanche, A. Metz, G. Pflug, J. F. Busch, V. C. Chappell, D. A. Sipfle, and M. Capek.

K53 Harig, Gerhard and Joseph Schleifstein, eds. Naturwissenschaft und Philosophie. Beitraege zum Internationalen Symposium ueber Naturwissenschaft und Philosophie Anlaesslich der 550-Jahr-Feier der Karl-Marx-Universitaet Leipzig. Berlin: Akademie-Verlag, 1960.

Includes: Gerhard Heber, "Ueber einige Philosophisch wichtige Aspekte der Quantentheorie," 27-32; Ernst Schmutzer, "Uber das Wesen und den Gehalt der Relativitaetstheorie," 33-45; Ch. M. Fataljiew, "Das Problem der Kausalitaet und die Moderne Physik," 47-59; Max von Laue, "Erkenntnistheorie und Relativitaetstheorie," 61-69.

K54 Heisenberg, W. Das Naturbild der heutigen Physik, Hamburg: Rohwolt, 1955.

K55 Heisenberg, Werner. Physics and Philosophy: The Revolution in Modern Science. New York: Harper & Row, 1958. 206 pp. Translation: Physik und Philosophie. Stuttgartt: S. Hirzel, 1959. 201 pp. Fisica e filosofia. Milano: Il Saggiatore, 1961. 213 pp.

Includes an introduction by F. S. C. Northrop, and reprints Heisenberg's essay in the development of quantum mechanics and its interpretation.

K56 Heisenberg, W. Mutamenti nelle basi della scienza, Torino: G. Einaudi, 1944. 106 pp.

K57 Heisenberg, Werner. Natural Law and the Structure of Matter. London: Rebel Press, 1970. 45pp.

Lecture given in 1964 in Athens, on the victory of Platonists over atomists.

K58 [Heisenberg, W.] Horz, Herbert. Werner Heisenberg und die Philosophie. Second edition. Berlin: Deutscher Verlag der Wissenschaften, 1968. 316 pp.

K59 Holton, Gerald. "Mach, Einstein, and the search for reality. Daedalus, 97 (1968): 636-673. Reprinted in item A46.

K60 Howson, Colin, ed. Method and Appraisal in the Physical Sciences. The critical background to modern science, 1800-1905. New York: Cambridge University Press, 1976. vii + 344pp.

Includes I. Lakatos, "History of Science and its Rational Reconstructions," and case studies supporting the Lakatos doctrine; and a critique by P. Feyerabend. Includes a reprint of item FA39.

K61 Jeans, J. H. The new background of science. New York: Macmillan Co.; Cambridge, Eng.: The University Press, 1933. viii + 301 pp. Translations: Die neuen Grundlagen der naturer Kenntnis. Stuttgart: Deustche Verlags-Anstalt, 1935. 340 pp. Les Nouvelles Bases Philosophiques de la Science. Paris: Hermann, 1935. 308 pp.

K62 Jeans, J. H. Physics and Philosophy. London: Cambridge University Press, 1942. 222 pp.

K63 Joravsky, David. Soviet Marxism and Natural Science 1917-1932. New York: Columbia University Press, 1961. xiv + 433 pp.

Chapter 18, The 'Crisis' in Physics.

K64 Kanitscheider, Bernulf. "Einstein's Treatment of Theoretical Concepts." Albert Einstein: His Influence on Physics, Philosophy and Politics (item BE9), pp. 137-58.

K65 Kedrov, B. M. Lenin I Revoliutsii͡a V Estestvoznanii XX Veka. Moskva: Nauka, 1969. 397 pp.

Lenin and the revolution in 20th century science.

K66 Kedrov, B. M. Lenin i Dialektika Estestvoznanii͡a XX Veka: Materii͡a i Dvizhenie. Moskva: Nauka, 1971. 399 pp.

Lenin and dialectic of nature of 20th century, Matter and Motion.

K67 Kedrov, B. M. "V. I. Lenin and the revolution in natural science at the turn of the 19th and 20th centuries." Acta Historiae Rerum Naturalium nec non technicarum, 14 (1981): 11-38.

K68 Klimaszewsky, Gunter. "Lenins philosophisches Erbe und die Entwicklung der modernen Physik." Einheit, 24 (1970): 308-18.

K69 Kuznetsov, B. G. "Lenin, Lanzheven i Predystoriia Teorii Otnositel'nosti." Voprosy Filosofii, 5 (1969): 24-27.

K70 [Lorentz, H. A.] McCormmach, Russell. "H. A. Lorentz and the electromagnetic view of nature." Isis, 61 (1970): 459-497.

K71 Miller, Arthur I. "On Lorentz's methodology." British Journal for the Philosophy of Science., 25 (1974): 29-45.

Critique of Zahar (item FA39).

K72 Malisoff, William Marias. "Physics: The decline of mechanism." Philosophy of Science, 7 (1940): 400-14.

K73 Margenau, Henry. The nature of physical reality. New York: McGraw-Hill, 1950. vii + 479p.

K74 Margenau, Henry. Thomism and the Physics of 1958: A confrontation. Milwaukee: Marquette University Press, 1958. 61 pp.

K75 Margenau, Henry. "The Method of Science and the Meaning of Reality." Integrative Principles of Modern Thought. Edited by Henry Margenau. New York: Gordon & Breach, 1976, pp. 3-43.

K76 Mercier, Andre. Stabilite, complementarite et determinabilite. Les fondements de la physique moderne. Lausanne: F. Rouge, 1942. 76 pp.

K77 Ne'eman, Yuval. "A view from the Bridge: Physics and Philosophy in the 20th Century." Connaissance Scientifique et Philosophie. Brussels: Palais des Academies, 1975, pp. 193-215.

Notes that Popper's falsifiability thesis has permeated our whole approach to physics. There is a discussion of particle physics since 1957, with brief remarks on space and time.

K78 Neurath, Otto. Le developpement du Circle de Vienne et L'avenir de l'empirisme logique. Paris: Hermann, 1935. Translation: II circolo di Vienna e l'avvenire dell'empirismo logico, Armando Armando, Editore. Roma, 1977. 111 pp.

K79 Nye, Mary Jo. "Gustave LeBon's Black Light: A study in physics and philosophy in France at the turn of the century.," *Historical Studies in the Physical Sciences*, 4 (1974): 163-95.

K80 Omel'ianovskii, M.E. *V. I. Lenin i fizika XX veka*. Moskva: Ogiz Gospolitizdat, 1947. 120 pp + errata. Translation: *Lenin es a XX szazao fizikaja*. Forditotta: Lovas Gyorgy. Budapest: "Szikra," 1949. 156 pp.

K81 [Pauli, W.] Hendry, John. "Pauli as philosopher." *British Journal for the Philosophy of Science*, 32 (1981): 277-82.

Essay review of *Wolfgang Pauli: Wissenschaftlichen Briefwechsel*, Bd. I.

K82 Pauli, W. *Fisica e conoscenza*. Torino: Boringhieri, 1964.

K83 [Planck, M.] Kretzschmar, Hermann. *Max Planck als Philosoph*. Muenchen: Reinhardt, 1967. 115 pp.

K84 [Planck, M.] Lewi, Sabina. "O przeciwfenomalistycznej postawie Maxi Plancka." *Kwartalnik Historii Nauki i Techniki*, 12 (1967): 331-40.

In Polish, with long English abstract. "On the antiphenomenalistic attitude of Max Planck."

K85 Planck. *La conoscenza del mondo fisico*. Torino: Einaudi, 1942. 413 pp.

K86 Poincaré, Henri. *La Science et l'Hypothese*. Paris: Flammarion, 1968. 252pp. Preface by Jules Vuillemin, 7-9. Translation: *Science and Hypothesis*. Authorized translation by George Bruce Halstead. With a special preface by H. Poincaré and an introduction by Prof. Josiah Royce. New York: The Science Press, 1905. xxxi + 198 pp.

K87 [Poincaré, H.] Paul, John P. *An analysis and evaluation of Henri Poincare's Cosmology, Epistemology, and Philosophy of Science*. Ph.D. Dissertation, Marquette University, 1969. 189 pp.

For summary see *Dissertation Abstracts International*, 31 (1970): 427-A.

K88 Prigogine, Ilya, and Isabelle Stengers. "The New Alliance, Part One -- From Dynamics to Thermodynamics: Physics, The Gradual Opening towards the World of Natural Processes." "...Part Two -- An Extended Dynamics: Towards a Human Science of Nature." *Scientia*, 112 (1977): 319-32, 643-53. Also published in French, *ibid*. 287-304, 617-30, and in Italian, *ibid*. 305-18, 631-41.

A survey of the implications of irreversible thermodynamics and quantum theory.

K89 Prigogine, Ilya. *La Nuova Alleanza. Uomo e natura in una scienza unificata*. Milano: Longanesi, 1979. 353 pp.

Preface by G. Toraldo di Francia; contains translations of various general and epistemological papers by I. Prigogine, Isabelle Stengers, Peter M. Allen, Robert Herman and others.

K90 Redondi, Pietro. *Epistemologia e storia della scienza. Le svolte teoriche da Duhem a Bachelard*. Milano: Feltrinelli, 1978. 256pp.

K91 Reichenbach, Hans. *L'analisi filosofica della conoscenza scientifica*. Padova, 1968.

K92 Rémy, Lestienne. "Unité et ambivalence du concept de temps physique." Paris: Centre National de la Recherche Scientifique, Centre de Documentation Sciences Humaines, 1978. 205 pp. (Cahiers d'histoire et de philosophie des sciences, 9).

K93 Rencontres internationales de Genève, Editions de la Baconnière, Neuchâtel, 1952-58. Italian translation: *Discussione sulla fisica moderna*, Torino: Boringhieri, 1959. 131 pp.

Essays by W. Heisenberg, Max Born, Erwin Schroedinger, Pierre Auger and discussions by Albert Picot, Giacomo Devoto, Ellen Juhnke, René Schaerer, Umberto Campagnolo, Daniel Christoff, Léon Rosenfeld, Arthur H. Compton.

K94 Rey, Abel. *La théorie de la physique chez les physiciens contemporains*. Paris: Alcan, 1930. 346 pp.

K95 Rosenthal-Schneider, Ilse. *Reality and Scientific Truth. Discussions with Einstein, von Laue, and Planck*. Detroit: Wayne State University Press, 1980. 150 pp.

K96 Schroedinger, Erwin. Scienza e Umanesimo. Che cos'é la vita. Firenza: Sansoni, 1970. 204 pp.

Contains translations of: "Science and humanism"; "Physics in our time"; "What is life"; "The physical aspect of the living cell".

K97 Somenzi, Vittorio. "La filosofia e la metodologia della scienza oggi in Italia." Man and World, 2 (1969): 285-95.

K98 Stebbing, L. Susan. Philosophy and the Physicists. New York: Dover Publications, 1958. xvi + 295 pp. (reprint of 1937 ed.)

Critique of the views of J. H. Jeans, A. Eddington and others.

K99 Stoll, Ivan. "Některe filozofické otázky současné fyziky." Nová mysl, 33, nr. 11 (1979): 43-154

Some philosophical questions of contemporary physics.

K100 Svyechnikov, G. A. Marxismo e causalita in fisica. Milano: Mazzotta, 1975. xiv + 256 pp. Translated from the English edition published by Progress Publishers, Moscow; the original Russian title is Prichinnost' i sviâz' sostoîanii v fizike.

K101 Tagliagambe, Silvano. Scienza, filosofia, politica in Unione Sovietica (1924-1939). Milano: Feltrinelli, 1978. 524 pp.

K102 Toulmin, Stephen, ed. Physical Reality. New York: Harper & Row, 1970. xx + 220 pp.

Reprints and translations of papers by Max Planck, Ernst Mach, T. Percy Nunn, Moritz Schlick, A. Einstein, B. Podolsky, N. Rosen, Niels Bohr, N. R. Hanson, P. K. Feyerabend, and David Bohm, on the interpretation of quantum mechanics and other physical theories.

K103 Ulehla, Ivan. "Vliv pozitivismu na ceskou fyziku." Pokroky matematiky, fyziky a astronomie, 24 (1979): 268-83.

"The influence of positivism on Czech physics."

K104 Úlehla, Ivan. "Vývoj fyziky a její vztah k filosofii." _Nova Mysl_, 27 (1973): 491-97.

"The development of physics and its relation to philosophy."

K105 Wallace, William A. _Causality and Scientific Explanation_. Volume Two. _Classical and Contemporary Science_. Ann Arbor: University of Michigan Press, 1974. xi + 422 pp.

K106 Weizsäcker, Carl Friedrich von. _Zum Weltbild der Physik_. 7th ed. Stuttgart: Hirzel, 1958. 378 pp.

K107 Weizsäcker, Carl Friedrich von. "Kant's first analogy of experience and conservation principles of physics." _Synthese_, 23 (1971): 75-95.

On the concept of substance.

K108 Wigner, Eugene P. "The unreasonable effectiveness of mathematics in the natural sciences." _Communications on Pure and Applied Mathematics_, 13 (1960): 1-14.

Includes remarks on the development of quantum theory.

K109 Yourgrau, Wolfgang and Allen D. Breck, eds. _Physics, Logic, and History, Based on the First International Colloquium held at the University of Denver, May 16-20, 1966_. New York: Plenum Press, 1970. xiv + 336 pp.

Includes: K. Popper, "A realist view of logic, physics, and history"; A. Mercier, "Knowledge and Physical Reality"; W. V. O. Quine, "Existence"; D. D. Ivanenko, "The problems of unifying cosmology with microphysics"; J. P. Vigier, "Possible internal subquantum motions of elementary particles"; G. Gamow, "The three kings of physics"; H. J. Treder, "Relativity theory and hisotricity of physical systems"; H. Bondi, "General Relativity as an open theory"; A. Lande, "The non-quantal foundations of quantum mechanics"; other papers and discussion.

K110 Zahar, Elie. "Mach, Einstein and the Rise of Modern Science." *British Journal for the Philosophy of Science*, 28 (1977): 195-213.

The general theory of relativity was compatible with Mach's philosophy but not the Special Theory. "Positivism was largely irrelevant to the development of modern physics." See critique by Feyerabend, item K46.

K111 Zahar, E. "Second thoughts about Machian positivism: A reply to Feyerabend." *British Journal for the Philosophy of Science*, 32 (1981): 267-76.

* * * * * *

For additional references see HW, sections W, F.f, L.i, and section B under the names of N. Bohr, B. Born, A. Einstein, W. Heisenberg, E. Schroedinger, etc.; see also *Isis CB*, "Personalities" volumes, under these names.

L CULTURAL INFLUENCES OF PHYSICS

L1 Adams, J. T. "Henry Adams and the new physics." *Yale Review*, 19 (1929): 283-302.

L2 Angrist, Stanley W., and Loren G. Hepler. "Demons, Poetry, and life: A thermodynamic view." *Texas Quarterly*, 10 (1967): 26-35.

L3 Arnheim, Rudolf. *Entropy and Art: An essay on order and disorder*. Berkeley: University of California Press, 1971. 64 pp.

L4 Bailey, John M. "Physics and everything: A bibliography." *American Journal of Physics*, 39 (1971): 1347-52.

A list of 203 items on relations of physics with other fields, for use by students in writing research papers.

L5 Beardsley, Monroe C. "Order and Disorder in Art." *The concept of Order*. Edited by Paul G. Kuntz. Seattle: University of Washington Press, 1968, pp. 191-218.

L6 Blass, Joseph. *Indeterminacy as a factor in scientific and artistic attitudes of the Twentieth Century*. Ph. D. Dissertation, Florida State University, 1968. 264 pp.

For summary see *Dissertation Abstracts International*, 29 (1968): 2298-A.

L7 Bork, A. M. "Randomness and the Twentieth Century." *Antioch Review*, 27 (1967): 40-61.

* Brush, Stephen G. "The chimerical cat: philosophy of quantum mechanics in historical perspective." Cited above as item HD9.

L8 Charbon, Remy. *Die Naturwissenschaften im modernen deutschen Drama*. Aurich: Artemis-Verlag, 1974. 282 pp.

On B. Brecht's *Leben des Galilei*, F. Dürrenmatt's *Die Physiker*, and H. Kipphardt's *In der Sache J. Robert Oppenheimer*.

L9 Corrington, J. W. "The new physics vs. the old Gnosis." *New Orleans Review*, 4 (1974): 99-106.

On T. Roszak's critique of physics.

L10 Duggan, I. Pius. *Relativity, Quantum Theory, and the novels of Samuel Beckett*. Ph. D. Dissertation, Loyola University of Chicago, 1971. 383 pp.

For summary see *Dissertation Abstracts International*, 32 (1971): 2637-A.

L11 Falk, Florence A. "Physics and the Theatre: Richard Foreman's *Particle Theory*." *Educational Theatre Journal*, 29 (1977): 395-404.

"Like the scientist, Foreman is intrigued by the question of matter itself. All of his works deal with the ambiguous nature of matter and with the question of whether phenomena have any underlying structural reality at all."

L12 Feinberg, E. L. "Vzaimosviaz' Nauki i Iskusstva v Mirovozzrenii Einshteina." *Voprosy Filosofii*, 3 (1979): 32-46. Also in *Einshteinovskii Sbornik 1977*, pp. 187-213.

Discusses connections of art and science in Einstein's thought.

L13 Friedman,, Alan J. "Physics and literature in this century: A new course." *Physics Education*, 8 (1973): 305-8.

L14 Friedman, Alan J., and Manfred Puetz. "Science as metaphor: Thomas Pynchon and *Gravity's Rainbow*." *Contemporary Literature*, 15 (1974): 345-59.

L15 Friedman, Alan J. "The novelist and modern physics: new metaphors for traditional themes." *Journal of College Science Teaching*, 4 (1975): 310-12.

L16 Georgescu-Roegen, N. *The Entropy Law and the Economic Process*. Cambridge, Mass. Harvard University Press, 1971. xvii + 458 pp.

L17 Holtz, William. "Thermodynamics and the Comic and Tragic Modes." *Western Humanities Review*, 25 (1971): 203-216.

Discusses the influence of the Second Law and the concept of interacting systems in modern literature, especially the works of H. Bergson.

L18 Hye, Allen E. "Bertolt Brecht and Atomic Physics." *Science/Technology & The Humanities*, 1 (1978): 157-170.

Argues that Brecht understood and applied the concept of statistical causality.

L19 Jaki, Stanley L. _The Relevance of Physics_. Chicago: University of Chicago Press, 1966. 604 pp.

Contents: I. The Chief World Models of Physics: The World as an Organism, as a Mechanism, and as a Pattern of Numbers. II. The Central Themes of Physical Research: The layers of matter, the frontiers of the cosmos, the edge of precision. III. Physics and other Disciplines: Biology, Metaphysics, Ethics, Theology. IV. Physics: Master or Servant? The fate of physics in scientism, and in human culture.

* Kipphardt, Heinar. _In the Matter of J. Robert Oppenheimer. A Play Freely Adapted on the Basis of the Documents_. Cited below as item JG13.

L21 Kuznetsov, B. G. _Einstein and Dostoyevsky. A Study of the relation of modern physics to the main ethical and aesthetical problems of the XIXth century_. London: Hutchinson Educational, 1972. 111 pp.

L22 Nadeau, Robert L. "Physics and Cosmology in the Fiction of Tom Robbins." _Critique_, 20 (1978): 63-74.

On the influence of quantum theory and relativity in Robbins' _Another Roadside Attraction_ (1971) and _Even Cowgirls get the Blues_ (1976).

L23 Nadeau, Robert. _Readings from the New Book on Nature: Physics and Metaphysics in the Modern Novel_. Amherst: University of Massachusetts Press, 1981. 213 pp.

Influence of ideas of quantum theory and relativity on John Fowles, John Barth, John Updike, Kurt Vonnegut, Jr., Thomas Pynchon, Tom Robbins, and Don DeLillo; general remarks on physics, metaphysics and the form of the novel.

L24 Raven, C. E. "The impact of physics on science and religion." _A Physics Anthology_. Edited by N. Clarke. London: Chapman & Hall, 1960, pp. 33-46. Reprinted from _Bulletin of the Institute of Physics_, July 1953.

L25 Restivo, Sal P. "Parallels and paradoxes in modern physics and Eastern mysticism: I. - A critical reconnaissance." _Social Studies of Science_, 8 (1978): 143-81.

L26 Restivo, Sal, and Michael Zenzen. "A humanistic perspective in science and society." _Humanity and Society_, 2, no. 4 (Nov. 1978): 211-36.

Includes comments on G. Chew's "bootstrap" theory and D. Bohm's "holonomy."

L27 Richardson, D. B. "Cultural influences in contemporary space theory." *Scientia*, 107 (1972): 575-92.

L28 Rifkin, Jeremy. *Entropy. A new world view*. New York: Viking, 1980. 305 pp.

L29 Ryan, Steven T. "Faulkner and Quantum Mechanics." *Western Humanities Review*, 33 (1979): 329-339.

"One can see in the metaphoric value of Caddy (in *The Sound and the Fury*) an attitude toward the life force which parallels the Bohr-Heisenberg view of the subatomic world."

L30 Schlant, Ernestine. "Hermann Broch and Modern Physics." *Germanic Review*, 53 (1978): 69-75.

Discusses how Broch came to know about relativity and quantum theory, which influenced his fiction according to Ziolkowski. See also item FD15.

L31 Siegel, Armand, and Mildred Siegel. "The accelerator and the Virgin: the rise and fall of two cults." *For Dirk Struik* (Boston Studies in the Philosophy of Science, 15) Edited by R. S. Cohen et al. Dordrecht: Reidel, 1974, pp. 403-15.

On physics after World War II.

L32 Simberloff, Daniel. "Entropy, information, and life: Biophysics in the novels of Thomas Pynchon." *Perspectives in Biology and Medicine*, 21 (1978): 617-25.

L33 Steinmetz, Joseph James. *Between Zero and One: A Psychohistoric Reading of Thomas Pynchon's Major Works*. Ph.D. Dissertation, University of Wisconsin, Madison, 1975. 336 pp. Summary in *Dissertation Abstracts International*, 37 (1976): 318-A.

Includes alleged influence of theories of Bohr, Heisenberg, and Pauli on *Gravity's Rainbow*.

L34 Tarbell, D. Stanley. "Perfectibility vs. Entropy in Recent Thought." *Science/Technology & The Humanities*, 1 (1978): 103-113.

Remarks on Henry Adams, E. Schroedinger, and N. Georgescu-Roegen.

L35 Waddington, D. H. *Behind appearance. A Study of the Relations between Painting and the Natural Sciences in this Century*. Cambridge, Mass.: MIT Press, 1970. 256 pp.

L36 Walter-Echols, Elizabeth. _Relativity and Totality: Science as Structure and Imagery in Selected Texts from Hermann Broch_. Ph.D. Dissertation, Indiana University, 1977. 279 pp. Summary in _Dissertation Abstracts International_, 38 (1977): 2154-A.

"Relativity theory, atomic theory and the principles of thermodynamics are the important but largely neglected sources for the primarily visual images in these works."

L37 Weinberg, Steven. "Reflections of a working scientist." _Daedalus_, 103, no. 3 (1974): 33-45.

Reply to critique of science by T. Roszak and others.

L38 Whyte, L. L. "Atomism, structure and form. A report on the natural philosophy of form." _Structure in Art and Science_. Edited by G. Kepes. New York: Braziller, 1965, pp. 20-28.

L39 Yukawa, H. _Creativity and Intuition: A Physicist Looks at East and West_. New York: Kodansha, 1973.

* * * * *

For additional references on this topic see HW, sections U.a & u.c.

M RESEARCH IN HISTORY OF MODERN PHYSICS AND ITS USE IN EDUCATION, ETC.

* Arons, A. B. "Phenomenology and logical reasoning in introductory physics courses." Cited above as item JB3.

M1 Boutry, G. A. "The historical approach in the teaching of physics." Why Teach Physics. Edited by S. C. Brown et al. Cambridge, Mass.: MIT Press, 1964, pp. 85-96.

M2 Bridgman, P. W. "Impertinent reflections on history of science." Philosophy of Science, 17 (1950): 63-73.

M3 Brown, Sanborn C., Donald Lindsay, and Stephen G. Brush. "The History of Science and its Place in a Physics Course." Teaching School Physics, A UNESCO Source Book. Edited by John L. Lewis. London: Taylor & Francis, 1972, pp. 122-33.

M4 Brush, Stephen G. "The role of history in the teaching of physics." Physics Teacher, 7 (1969): 271-80

Illustrates the possible educational value of history by discussing three cases: (1) the first estimate of the speed of light (not by O. Roemer!): (2) Kelvin and energy dissipation in connection with the cooling of the earth; (3) Planck's quantum theory and the myth of the "ultraviolet catastrophe."

M5 Brush, Stephen G., and Allen L. King, eds. History in the Teaching of Physics. Proceedings of the International Working Seminar on the Role of the History of Physics in Physics Education. Hanover, N.H.: University Press of New England, 1972. xi + 116 pp.

Includes M. J. Klein, "The use and abuse of historical teaching in physics;" G. Weiner, "Resource centers and programs for the history of physics;" S. C. Brown, "Organization of international efforts in using history of physics in physics teaching;" reports of working groups and recommendations.

M6 Brush, Stephen G., ed. Resources for the History of Physics. Hanover, NH: University Press of New Hampshire, 1972. 86 + 90 pp.

Part I: Guide to books and audiovisual materials. Part II: Guide to Original Works of Historical Importance and their translations into other languages.

M7 Brush, Stephen G. "Should the History of Science be rated X?" Science, 183 (1974): 1164-72.

M8 Caldirola, Piero, and Angelo Loinger. "Storie e filosofie della fisica." _Epistemologia_, 1 (1978): 77-92. Also in Caldirola and Loinger, _Teoria fisica e realta_. Napoli: Liguori, 1979, pp. 148-162.

M9 Davis, Harold L. "Committing ourselves to study the past." _Physics Today_, 33, no. 6 (June 1980): 104.

On the formation of the History of Physics Division in the American Physical Society.

M10 Delokarov, K. Kh. "S. F. Vasil'ev i razrabotka metodologicheskikh problem razvitiia fizik." _Voprosy Istorii Estestvoznaniia i Tekhniki_, 4 (1981): 52-60.

On S. F. Vasilev's work on methodological problems of physics history, especially concerning quantum mechanics and relativity, in the 1920s and 1930s.

M11 Devons, Samuel, and Lillian Hartmann. "A history-of-physics laboratory." _Physics Today_, 23, no. 2 (Feb. 1970): 44-49.

M12 Devons, Samuel. "History of Physics Laboratories." _AAPT Announcer_, 8, no. 4 (Dec. 1978): 44-45. (American Association of Physics Teachers)

List of 12 laboratories at various colleges, some with brief descriptions.

* Diederich, Mary E. "The context of inquiry in physics." Cited above as item K32.

M13 Dorfman, Ia. G. and O. A. Lezhneva. "Istoriîa Fiziki." _Voprosy Istorii Estestvoznaniîa i Tekhniki_. 23 (1968): 32-36.

On Soviet historiography of physics and history of physics research 1918-1968.

M14 [Einstein, A.] Frankfurt, U. I. and A. M. Frenk. "Einshtein kak Istorik Nauki." _Einshteinovskii Sbornik_. Moskva: Nauka, 1966, pp. 298-338.

Einstein as an historian of science.

M15 Frenkel' V. Ia. "O vospriimchivosti k novym fizicheskim ideiam i rezul'tatam." _Nauchnoe otkrytie i ego vospriiatie_. Moskva: Nauka, 1971, pp. 296-307. Also in _Ideen des exakten Wissens_, 9 (1969): 527-534.

M16 Frenkel' V. Ia. "Wissenschaftliche Biographien." _Ideen des exakten Wissens_, 2 (1972): 85-92. Translation: " O zhanre biografiĭ uchenyk." _Chelovek Nauki_. Moskva: Nauka, 1974, pp. 108-124.

M17 Frenkel' V. Ia. "Izuchenie diskussiĭ v fizike - osnova ikh planirovaniiâ i provedeniiâ." Voprosy filosofii, 8 (1978): 94-98.

M18 Frenkel' V. Ia. "O diskussii i ee nauchnykh sekundantakh." Voprosy filosofii, 1 (1979): 112-115.

M19 Grigorian, A. T., ed. Istoriia Estestvoznaniia. Literatura, Opublikovannaia v SSSR (1967-1970). Moskva: Nauka, 1981. 768 pp.

Russian publications on history of science; physics section compiled by O. A. Lezhneva.

M20 Grote, Michael. "The history of physics in a high school physics course." Physics Teacher, 15 (1977): 102-3.

M21 Gruenbaum, A. "The bearing of philosophy on the history of science." Science, 143 (1964): 1406-12.

Includes remarks on the history of special relativity.

M22 Gulo, D. D. "Estestvennonauchnye i Filosofskie Vzgliady N. A. Umova." Istoriîa i Metodologiîa Estestvennykh Nauk, 1 (1960): Fizika, 115-140.

On the scientific and philosophical views of historian of science N.A. Umov (with bibliography).

M23 Haas, Arthur Eric. "Der Wert der geschichtlichen Methode in physikalischen Unterricht." Zeitschrift fur mathematischen und naturwissen-schaftlichen Unterricht, 45 (1914): 281-86.

M24 Hermann, A. "Quellenfuer die Geschichtsschreibung der modernen Physik." Technikgeschichte in Einzeldarstellungen, Nr. 17. Duesseldorf, 1969, pp. 65-69.

M25 Hermann, A. "Begleitwort" (ueber Probleme und Aufgaben der Wissenschaftsgeschichte). Geschichte der mechanischen Prinzipien, by Istvan Szabo. Sonderausgabe der Stiftung Volkswagenwerk, Basel und Stuttgart, 1976, pp. xi-xx.

M26 Hermann, A. "Das Verhaeltnis von Naturwissenschaft und Technik in historischer Sicht." Technikgeschichte, 43 (1976), 116-124.

M27 Hermann, A. "Die Funktion und Bedeutung von Briefen." Wolfgang Pauli, Wissenschaftlicher Briefwechsel mit Bohr, Einstein, Heisenberg u.a., Bd. I: 1919-1929. New York-Heidelberg-Berlin: Springer-Verlag, 1979, pp. xi-xlvii.

M28 Hermann, A. "Technik und Natur." Funkkolleg Geschichte. Studienbegleitbrief 3, Hrsg. DIFF. Tubingen, Weinheim und Basel, 1979, pp. 49-77.

M29 Herman, A. "Die Technik als Kulturfaktor." Kultur & Technik, 4, no. 1 (1980): 1-5. Also in Die Technikgeschichte als Vorbild moderner Technik. Schriften der George - Agricola - Gesellschaft Nr. 6 (1980), pp. 22-27.

M30 Hirosige, Tetu. "Studies of History of Physics in Japan." Japanese Studies in History of Science, 1 (1962): 26-34.

M31 Hirosige, Tetu. "Activities of Japan's Group for History of Physics." Japanese Studies in the History of Science, 9 (1970): 5-12.

M32 Holton, Gerald. "A new look at the historical analysis of modern physics." Actes XIIIe Congres International d'Histoire des Sciences, Conferences Plenieres, 1971 (pub. 1974): 88-144.

Includes remarks on the unification of science, and of scientific and social concerns; on different approaches to the history of science; and on the work of Albert Einstein.

M33 Hulin, Nicole, and Danielle Fauque. "History of Science and Physics Teaching." CUIDE (Universite Paris), no. 21 (February 1982).

Describes an experimental physics course including historical material on falling bodies and optics; includes bibliography of sources in the history of physics for both teachers and pupils.

M34 Illizarov, S. S. and O. A. Lezhneva, compilers; Fedoseev, I. A., ed. Institute of the History of Natural Sciences and Technology (Brief Review). Moscow: Nauka, 1981. 139 pp.

Gives information on the organization of research in the USSR in the history of science and technology; includes a bibliography.

M35 Johnson, E. H. "Courses on the history of physics in American colleges and Universities." Science, 74 (1931): 435-37.

M36 Kaga, Martin H., and Eric Mendoza. "Down with the history of relativity!" Physics Teacher, 16 (1978): 225-27.

M37 Kaminer, L. V.; L. Ia. Pavlova; P. V. Pil'shchikova; A. T. Grigor'ian; D. D. Ivanov, eds. _Istoriia Estestvoznaniia. Literatura opublikovannaia v SSSR (1957-1961)_ Moskva: Nauka, 1972. 222 pp.

Bibliography of history of science literature published in the USSR in 1957-61.

M38 Kondo, Herbert. "History and the Physicist." _American Journal of Physics_, 23 (1955): 430-36.

M39 Krafft, Fritz. "Physikgeschichte fuer den Unterricht." _Technikgeschichte_, 41 (1974): 341-51.

Review of A. Hermann (ed.) _Lexikon Geschichte der Physik A-Z_ (Koeln: Aulis Verlag Deubner & Co. KG, 1972).

M40 Krafft, Fritz. "Naturwissenschaftsgeschichte in Lehre und Forschung." _Physikalische Blaetter_, 31 (1975): 385-95.

M41 Krafft, Fritz. _Naturwissenschafts- und Technikgeschichte in der Bundesrepublik Deutschland und in West-Berlin 1970-1980. Eine Uebersicht ueber die Forschung und Lehre an den Institutionen_. Wiesbaden: Akademische Verlagsgesellschaft Athenaion, 1981. (Also: _Berichte zur Wissenschaftsgeschichte_, Sonderheft) 163 pp.

Includes bibliography of publications by historians of physics, H. Kangro, A. Kleinert, U. Hoyer, W. Gerlach, F. Fraunberger, A. Hermann, D. C. Cassidy, K. v Meyenn, J. G. O'Hara.

M42 Krans, Ruud L. "The history of physics in the education of physics teachers." _Physics Education_, 7 (1972): 58-60.

M43 [Kravets, T.] Faerman, G. P. "Torichan Pavlovich Kravets." _50 Let Gosudarstvennogo Opticheskogo Instituta im S. I. Vavilova 1918-1968_. Leningrad (1968), pp. 668-683.

On historian of physics T. P. Kravets (1876-1955).

M44 Kuznetsov, V. I. "Reshenie fundamental'nykh problem estestvoznaniia istorikonauchnymi metodami." _Voprosy Istorii Estestvoznaniia i Tekhniki_, 1, no. 34, (1971): 42-46.

The solution of fundamental problems of science by the methods of history of science.

M45 Leacock, R. A. and H. I. Sharlin. "The nature of physics and history: A cross-disciplinary inquiry." *American Journal of Physics*, 45 (1977): 146-53.

Description of a college course.

M46 Lezhneva, O. A. "Issledovaniia po Istorii Fiziki." *Razvitie Fiziki v SSSR*. Vol. 2. Nauka: Moskva, 1967, pp. 329-353.

M47 Lezhneva, O. A. "Istoriia Fiziki." *Voprosy Istorii Estestvoznaniia i Tekhniki*, 59, no. 2 (1977): 46-51.

On research in the history of physics in the USSR during the past 50 years.

M48 *Papers by Soviet scientists*. Moscow: Akademiya Nauk SSSR. Institut Istorii Yestestvoznaniya i Tekhniki, 1977. (Distributed at the International Congress of the History of Science, XIV (1977).)

M49 Peck, R. A., Jr. and W. E. Haisley, Jr. "A one-semester physics course for liberal arts students." *American Journal of Physics*, 23 (1955): 440-49.

Based on the text by G. Holton.

M50 Prochazkova, Eva and J. Folta. "Dejiny fyziky jako soucast vyuky budoucich ucitelu fyziky." *Matematika a fyzika ve skole*, 10 (1979-80): 139-46.

"The history of physics as a part of the educationof future teachers of physics."

M51 Prochazkova, Eva. "Zum Unterricht der Geschichte der Physik an den Mittelschulen und Hochschulen zur Heranbildung von Lehrern." *Acta Historiae Rerum Naturalium nec nonTechnicarum*, 11 (1981): 299-314.

M52 Rosmorduc, Jean. *L'Histoire et la Philosophie des Sciences et des Techniques en France. Chercheurs et Enseignements (Annuaire 1979-1980)*. Cahiers d'Histoire et de Philosophie des Sciences, Numero special. Paris: Centre de Documentation des Sciences Humaines du CNRS, 1980. 252 pp.

List of historians of physics with their fields of interest and publications, pp. 82-92.

M53 Schwippel, Jindrich. "Dokumente zur Geschichte der Wissenschaft und der Technik -- ihre gegenseitigen Beziehungen vom Gesichtspunkt des Archivwesens und vom Gesichtspunkt der Quellenkunde." *IV. Konferenz der Akademie-Archive* Sozialistischer Lander, Materialen, Ustredni archiv Ceskoslovenska Akademie Ved, Praha, 1976, pp. 128-71.

M54 Seeger, R. J. "On teaching the history of physics." *Journal of Physics*, 32 (1964): 619-25.

M55 Strugalski, Zbigniew. "Teaching of the history of physics." *Acta Historiae Rerum Naturalium nec non Technicarum*, 11 (1981): 287-98.

* Swenson, L. S., Jr. "The Michelson-Morley-Miller experiments and the Einsteinian synthesis." Cited above as item FA35.

* Truesdell, C. "Reactions of the History of Mechanics upon modern research." Cited above as item D9.

M56 Warnow, Joan N. "The physicist's research notes as historical documents." *Physics Today*, 24, no. 10 (Oct. 1971): 9.

M57 Warren, J. W. "The Development of the Teaching of Physics in Britain." *European Journal of Physics*, 1 (1980): 124-127.

"Physics was formerly learned in Britain through the study of the history of the subject. In recent times this approach has been found to be no longer possible, but there has been no systematic adoption of an alternative approach. In consequence there is much confusion and much incorrect teaching..." (From the author's abstract).

M58 Weart, Spencer. "Remembering ourselves." *Physics Today*, 31, no. 3 (March 1978): 120.

M59 Woodall, A. J. "Science history. The place of the history of science in science teaching." *Physics Education*, 2 (1967): 297-305.

NAME INDEX

SUBJECT INDEX

INSTITUTIONAL INDEX